THE RB211 – ITS ROLE IN THE FALL AND RISE OF ROLLS-ROYCE LTD.

THE 1971 BANKRUPTCY EXPLORED

THE RB211 – ITS ROLE IN THE FALL AND RISE OF ROLLS-ROYCE LTD.

THE 1971 BANKRUPTCY EXPLORED

Alan Edge

The views, opinions and commentaries expressed in this book are those of the author and individual contributors as well as officers of the companies at the time and are not attributable to Rolls-Royce Ltd., or its successor companies, Lockheed Aircraft Corporation and its successor Lockheed Martin. Every effort has been made on the part of the author to verify information and photographs where possible and copyright and acknowledgements have been duly given.

The various references to 'Rolls-Royce' should be understood to mean successor companies throughout the periods pre & post 1971.

I fully acknowledge any previously documented history and events of the Rolls-Royce receivership period, wherever they may exist.

Published by The Publishing Edge

A CIP catalogue record for this book is available from the British Library.

ISBN 978-1-0687999-0-7

Typeset and cover design by Clare Brayshaw

Front cover images: British Airways TriStar with RB211 engines, courtesy of British Airways Heritage Collection; RB211-22B engine fan from displayed engine, courtesy of Rolls-Royce Heritage Trust Exhibition, Derby.

Manufacturer: York Publishing Services Ltd
64 Hallfield Road, Layerthorpe, York YO31 7ZQ
Tel: 01904 431213 | Email: enquiries@yps-publishing.co.uk
Website: www.yps-publishing.co.uk

Represented by: Authorised Rep Compliance Ltd.
Ground Floor, 71 Lower Baggot Street, Dublin D02 P593, Ireland
www.arccompliance.com

CONTENTS

"There is no organisation, I almost said institution, in any country which compares with the firm of Rolls-Royce. No firm has such a solid background of technical achievement, employs so many brilliant specialists, and has such a fund of hidden resources upon which it can draw at will.

Whether in the making of engines for Bristol Fighters in World War I, in producing Schneider Trophy engines, or engines that powered the aircraft which won the Battle of Britain, Rolls-Royce has always lived up to the almost mythical standards of quality attaching to its name.

It has always followed a policy of going ahead with the development of projects which it considered to be essential."

Grierson, J. Test pilot, aviation pioneer and author
Jet Flight. Sampson, Low, Manston & Co. Ltd. (1945)

THE AUTHOR

Rolls-Royce Commercial Department 1988 – Author is located Second Row third left

The Author, Alan Edge, joined Rolls-Royce Limited in Derby in August 1970, as a commercial trainee, six months before the company went into receivership. He went on to enjoy a varied procurement and commercially oriented career at Rolls-Royce and also briefly at Rolls-Royce Canada. Over a 20-year period he gained considerable expertise in negotiating and managing customer engine contracts worldwide, finally leaving his permanent position in 1991, although returning later as a contractor. During his Rolls-Royce career he was involved, from a commercial and professional point of view, with most of the company's civil aero engine portfolio of the 1970's and 1980's. These included the Dart, Spey, and Tay and, of course, the RB211 engine across all its -22B, -524 and -535 series variants.

After leaving Rolls-Royce he went on to continue his commercial career working for other blue chip companies including GEC Alstom, BAE Systems, National Grid, Bombardier Transportation and Transpower, in New Zealand, on transportation and power procurement projects worldwide.

He also gained an M.A. degree in Energy Studies. He is a member of the Royal Aeronautical Society and has continued to pursue and develop his interests in aviation history further by writing this, his first book, whilst also becoming an active member of the Rolls-Royce Heritage Trust.

FOREWORD

By Phil Ruffles

Phil Ruffles, CBE, FREng, FRS
(Former Rolls-Royce Ltd., Director of Engineering and Technology)

The decision to pursue the RB211 programme in 1967 was the most important decision taken in the long history of Rolls-Royce. On the one hand it resulted in the disastrous receivership on February 4th, 1971, but on the other hand it provided the foundations on which the company grew and prospered for the next 50 years.

This book describes the events leading up to and following the receivership from many different angles. The story is told through individual contributions from the people involved at all levels within the company, from the customers, from governments, from local politicians and from the trade unions. It is a unique way of telling the story and provides an insight into the events that took place to ensure the continuation of the programme and the ultimate success of Rolls-Royce.

Alan Edge is to be congratulated on producing such a remarkable book.

PREFACE AND DEDICATION

Prior to the recent catastrophic impact on the world aviation industry of the COVID-19 virus, few events in the long history of Rolls-Royce and all of its successor company formations have been so important, so traumatic and, in my opinion, potentially fateful, as the bankruptcy (receivership) of the company on February 4th, 1971. This left a lasting impression not only on me, as a sixteen-year-old Rolls-Royce trainee, but also, I believe, on those former employees, the citizens of Derby and at the time, the country as a whole.

In recent years the event resurfaced in my mind with a desire to explore and record this event of 1971, in a book particularly as the fiftieth anniversary of the Rolls-Royce receivership loomed in 2021. It's not intended to be a comprehensive blow-by-blow historical account, although there are a couple of detailed chronologies of key events. I acknowledge there have been other more detailed technical accounts written of the receivership period. However, it occurred to me and seemed only fitting that a more narrative account was long overdue, which included capturing the stories of that period; the fate of the RB211 aero engine programme, before and after the receivership; memories, stories and experiences from those involved across the workforce; those directly involved in the RB211 contract itself and with the receivership team, from former employees, customers and other key people, before that first-hand information was lost forever, given the current age of most of the contributors. I have also attempted to highlight some historical and social background as well as the corporate and the political landscape of the time. It also encompasses the main causes leading to the financial collapse along with some possible secondary influencing factors that led to the company being placed into receivership.

This seismic event and the aftershocks of being placed into receivership had a major effect and impact on the psyche of the company and remained with its employees for many years afterwards. It also greatly influenced the thinking and future direction of the company, eventually restoring its reputation and re-establishing it to a world leading position in aero engine manufacture. Additionally, as my research progressed, notwithstanding the sheer hard work and valiant efforts of the directors and employees, it was apparent to me that the work of the receiver, E. Rupert Nicholson and his team had, perhaps not been given the full credit they deserved for their efforts and key decision making to assist the rescue of the company and the saving of the RB211 engine programme with Lockheed Aircraft Corporation, thus securing the immediate future of Rolls-Royce.

Over fifty years have now passed since that extraordinary day in February 1971, which has made this a very challenging but also an extremely rewarding piece of research work whilst developing many personal friendships along the way. It hasn't been an easy research project, despite some good source material being available, it has been very difficult to find and locate some of those key people to interview given the passage of time. What should not be understated of course is that the number of responders would have been very much higher but for this and also the restrictions placed on potential contributors who were subject to confidentiality and non-disclosure agreement restrictions, which could not be accessed or utilised for this book.

Finally, I would like to recognise all former Rolls-Royce employees, management and directors and trade union leaders from that time who, along with Derby Town Council representatives and local Members of Parliament for Derby and Derbyshire who with a combined effort, fought for the survival and eventual preservation of Rolls-Royce and the RB211 engine programme.

I dedicate this book to my parents Nancy and George who helped guide me to become the person I am today.

INTRODUCTION

THE 4th OF FEBRUARY 1971

AFTER NEARLY 65 YEARS OF BUSINESS TRADING, ROLLS-ROYCE LIMITED, ARGUABLY ONE OF IF NOT THE MOST FAMOUS AND PRESTIGIOUS ENGINEERING COMPANIES IN THE WORLD, WAS DECLARED BANKRUPT ON THE 4th OF FEBRUARY 1971, AND PLACED INTO RECEIVERSHIP.

ROLLS-ROYCE

"By the late 1960s, the only airframe manufacturers able to design, build and sell the new generation of wide-body passenger and freight aircraft were the American firms Lockheed, Douglas and Boeing. Similarly, there remained only Pratt & Whitney (P&W), General Electric in America and Rolls-Royce in the United Kingdom which were capable of providing the new engines to power those aircraft. Rolls-Royce won the contract to supply its RB211 turbofan engine for the Lockheed L-1011 TriStar but under-estimated the costs and the time it would take to do so. As a result, Rolls-Royce was bankrupted, the financial markets being unwilling to take the risk of finding the huge sums required to overcome the problems of design and metallurgy facing the company. Reluctantly Mr. Heath's Conservative government took it into state ownership. The Treasury was no happier at being required to find the cash needed, but there was a political and industrial imperative to save the company and the jobs it provided."

The Rt Hon. the Lord Tebbit CH, PC (2018)
(Former RAF pilot and Conservative politician)

"In early 1971, Rolls-Royce Limited was in trouble over its contract with Lockheed. On the financial front things went from bad to worse. On a really successful programme, the money paid out on development and getting into production takes something like 15 years for the programme to move from the red into the black.

> *At the fourth or fifth years many millions of pounds have to be spent on production tooling, raw materials and finished parts which begin to fill the stores with 'work in progress'. If there is any minor hiccup in development, production is thrown into confusion, making engines and parts that cannot be delivered to the customer, while the stores fill up with costly parts that cannot be delivered to the customer.*
>
> *By mid-1970 it was clear that Rolls-Royce Derby was deep into financial difficulties and would be unable to fulfil its contractual obligations with Lockheed on the RB211. The company's enormous assets were no help because what it was fast running out of was cash and credit to pay the wages and the bills of the thousands of subcontractors and suppliers."*
>
> **Sir Stanley Hooker (1984) *"Not Much of an Engineer": An Autobiography, Airlife Publishing.***
> **(Rolls-Royce Ltd., Technical Director in 1971)**

This is the story of the key events and dramas surrounding and leading up to the placement of Rolls-Royce into official receivership on the 4th of February 1971. It sets out to record and capture this significant period of the social history of what was, still is and will remain, this great company, Rolls-Royce.

It also considers the company's eventual financial recovery after receivership and its future growth into a successful global business, leading onto its re-privatisation as a company in 1987. This ushered in the dawn of a new era of success for Rolls-Royce in the launch of the Trent family of engines, therefore embracing the company mantra first laid down by Sir Henry Royce of "Invent once, use many times". Finally, it considers the impact and stated nature of the company's future business following the 'black swan' event of the COVID-19 pandemic and notes some of the projects that will take this iconic company forward into the future.

The financial collapse of any company is a tragic loss and concern, not only for the company affected but for its employees and the local community alike. In the case of a company with the fame and prestige of Rolls-Royce, with its renowned, worldwide reputation for excellence in all things engineering, the financial crash was of national interest, and which could have been a devastating loss for Derby and the whole country had the company not survived.

The book contains many previously unpublished photographic images, documents and key commentaries that have been widely sourced and kindly loaned to help paint a picture, a snapshot if you like, of Rolls-Royce back then. Included is some technical information for the engineers that may read this, kindly supplied by those directly involved with the engine, leading up to, during and after this critical time in the company's history.

Hopefully the reader will gain some knowledge and insight into the developing financial crisis in which Rolls-Royce found itself engulfed during the late 1960's and early 1970's as it struggled to meet the challenges of developing a new engine and in achieving the RB211 engine delivery dates against a backdrop of escalating costs and decreasing cash liquidity.

CHAPTER 1

IN THE BEGINNING – ROLLS-ROYCE LIMITED EARLY HISTORY

THE FOUNDERS AND COMPANY FORMATION

CHARLES STEWART ROLLS

Born into the aristocracy in 1877, Charles Rolls' upbringing was completely opposite to that of Henry Royce, enjoying the comforts and privileges as would be customary and fitting at that time for the third son of Lord Llangattock. From a very early age he had shown great interest in all things mechanical and particularly vehicles of movement and great speed. He was also very keen on cycling, and he actually designed a quadruplet bicycle for four people. By his university days at Cambridge, he had already been introduced to early motor vehicles and, on gaining a Mechanical Engineering degree, he continued his passion for speed and competing in motor racing time trials.

Charles Rolls also developed an enthusiasm, desire and interest in the early days of the development of the 'flying machine'. He was stimulated by his 1907 meeting with Wilbur Wright, which probably led to many trips in hot air balloons, which piqued his interest in powered flight. It was this interest that was to lead to the tragic and abrupt ending of his life on July 12th, 1910. The plane he was flying in crashed, killing him instantly and thus ending the blossoming partnership with Henry Royce and their company. The ultimate irony of his death was truly emphasised when the ill-fated plane was photographed being towed away by a Rolls-Royce Silver Ghost motor car, perhaps truly fitting of such a great man and entrepreneurial adventurer.

FREDERICK HENRY ROYCE

Henry Royce, born in 1863, was a typical example of the archetypal career engineer. Born of humble beginnings, very early in his life he showed a strong interest in engineering and mechanical pursuits. His introduction to the world of engineering was being apprenticed to the Great Northern Railway works at Peterborough. However, to his great misfortune, the apprenticeship ended abruptly when his aunt ran short of money to fund his training, so one can only imagine the success he may have had, had he continued on into a railway engineering career, but the railway's loss was to be aviation's gain.

Royce then became involved in the early days of electricity generation when he took a position as a tester for the city of London Electric Light and Power Company. Later, in 1884, he went on to form F. H. Royce & Company of Cooke Street, Manchester, within which he sought to explore and develop new electro-mechanical products involving his newly found knowledge of electricity generation.

Around 1913, Henry Royce suffered deteriorating health problems and was unable to carry out his duties any more in Derby. He moved to the south coast to convalesce and operated from there as his base, until his death in 1933.

Effectively within seven years from the company's formation, Rolls-Royce had lost its founding fathers as a driving force and the burden of responsibility for the future direction and management of the company was left in the very capable hands of Claude Johnson who had been influential and a key figure in the early days of F. H. Royce & Co.

ROLLS-ROYCE LTD: THE EARLY YEARS – CARS TO PLANES, GHOSTS TO BIRDS

Rolls-Royce Limited was formed on March 16th, 1906, after an initial meeting of Rolls and Royce at the Midland Hotel, Manchester in May 1904. It had a start-up capital of £60,000. Soon after the company's formation it became evident that it would be impossible to continue production of electrical products at the Cooke Street, Manchester, works owing to insufficient space for the continued development and expansion of manufacturing that was taking place.

Following detailed investigations, it was finally decided to move the business from Manchester to Derby which at that time was offering cheaper rates of electrical power and had a greenfield 12-acre site that was available for development. This new factory at Nightingale Road was officially opened in July 1908. This had involved a considerable increase in the capital requirements of the company.

Manufacture of early models of motor cars was the key activity at the Derby site and production of probably the company's most famous and iconic model, the Silver Ghost, was in full swing around the outbreak of World War I in 1914. Rolls-Royce also developed and manufactured armoured cars, commencing production in 1914 which had been a government requirement arising as a result of the needs of World War I. It is reported that seventy-two were sent into battle in 1915. In addition, the onset of World War I had prompted the government to ask Rolls-Royce to produce aero engines of French design, which was not to the liking of Henry Royce. He immediately implemented a plan for the company to produce its own engines. Following on from these initiatives were developments of the first Rolls-Royce piston engine, the Eagle, later powering the Fairey 111C plane (float and wheel-based) and was the greatest production type used in the Vickers Vimy bomber aircraft, which later powered the first transatlantic flight in 1919. These new engine designs and products represented the very early days of aero engine development and manufacture by Rolls-Royce.

Rolls-Royce's business experienced mixed fortunes after World War I with demand for engines from the RAF vastly reduced and manufacture of the Eagle and Condor engines reaching maturity. However, there were two high spots for the company in 1929 and1931 when the Government Air Ministry decided to compete and win in the Schneider Trophy contest. Firstly, with a Supermarine S.6 and then S.6B aircraft both powered by the Rolls-Royce R(racing)engine developed from the Buzzard (H) engine, itself a derivative of the Kestrel engine.

It would not be until 1942 that Rolls-Royce was to make its next step change in technology and products with its involvement in the development of the jet engine with Sir Frank Whittle. This decision proved to be the catalyst for the future success of the company in the years to follow.

ROLLS-ROYCE MOVES INTO THE JET AGE

> *"The unlikely setting of the 'Swan and Royal Hotel', in Clitheroe, Lancashire, proved to be the location for a meeting and dinner in early December 1942, involving the leading executives of Rolls-Royce Limited and the Rover Car Company. The background to this event involved the commitment by Ernest Hives (later Lord Hives, when Chairman), the Chief Executive of Rolls-Royce to establish Rolls-Royce in the emerging technology of jet engines. At the time Frank Whittle was at the early stage of demonstrating the potential of his proposed jet engine in trials at a disused factory in Lutterworth, facing frequent failures on the test bed. The Chief Engineer at Rolls-Royce was Stanley Hooker and he and Hives visited Lutterworth to see the work in progress. After one such visit by Stanley Hooker alone, on returning to Derby he said to Hives that "if the potential of Whittle's ideas came to fruition the jet engine would exceed the power output of the latest Merlin piston engine" then the jewel in the crown of Rolls-Royce.*

World War II had begun, he had moved to 'war work' at Rolls-Royce and, as so many of this generation, enjoyed reasonable pay for the first time in his life.

The Internal Combustion Turbine Test Department

In early 1944, I was invited to go for an interview with Frank Greaves in the 'Technical Block' on Nightingale Road. Frank Greaves (FG) had arrived from Fairey Aviation in 1941 to be appointed as Rolls-Royce's first gas turbine engine tester. He was the third member of the team forming the Internal Combustion Turbine (ICT) department. These young men had good reason to be grateful for this pioneering experience in the gas turbine field, and I joined the ICT, based on Sinfin A site, on March 3rd, 1944. I was assigned various tasks during the early days including involvement in aerodynamic cascade testing, combustion chamber testing, compressor testing and engine testing. Life was very good for me and, in addition, most of us were attending college on a day-release basis.

Early Vibration Measurements

The mechanical balance of the engine through the speed range was a matter of some concern particularly when trying to assess whether flight engines were acceptable. The normal assessment method in the early days was by a 'laying on of hands'. The test inspector would stand alongside the engine as it was slowly accelerated over the speed range, and he decided whether he considered the vibration to be acceptable – very much a matter of opinion. Vibration meters (accelerometers) were gradually developed under the leadership of the head of the Vibration Department, a large man in every sense, named Dennis Drew (nicknamed affectionately as 'Dizzy' Drew). I remember on one occasion we had an early vibration meter mounted on the front of the compressor casing on an Avon engine and Dizzy decided to go and lay hands on it with the engine running at high speed.

I was horrified to see him stride straight up to the engine intake, obviously concentrating on what he was about to do, and of course his bulk restricted the airflow somewhat and the engine protested with a violent surge. We shut the engine down as quickly as possible. Dizzy emerged from the test-bed cell with great dignity and with his very nice double-breasted suit shredded in tatters. His only comment was that he would go and change his suit!

'Dizzy' Drew – The Perils of Engine Testing. Cartoon by Graham Newton

Altitude Test Facility at Sinfin 'C' Site

As a result of a reorganisation in May 1962, I moved to the Altitude Test Facility (ATF) as Test Superintendent where my previous experience in America came in very useful. Apart from the actual test operation, the ATF had been largely staffed by ex-service personnel because of an instruction by Lord Hives that existing Rolls-Royce people should not be used in supervisory positions (apart from the test superintendent job). As a result, the operation of the facility was conducted in quite a formal way with Group Captain George Watt, AFC, as General Manager, with ex-Navy Commanders and an ex-RAF Wing Commander heading the various sections. However, it has to be said that this system worked well because of the complexity of the operation and the relatively primitive state of instrumentation in those days.

External Test Facilities

In 1962, Frank Greaves had been appointed manager of External Facilities and, with various licensing arrangements, coupled later with the merger with the Bristol Siddeley Engine Group, this work

had expanded. In 1965 I moved again to spend an interesting two years following up of test facility investigations and conducting feasibility or viability studies on various proposed test facilities.

In 1967 I was appointed as Frank's deputy to assist in this work. In particular, I had the task of arranging for testing which could not be accommodated at the Sinfin sites and had to be carried out at various sites including Ansty, Brockworth, Dowty Gloucester, and the Patchway site, Bristol. A colleague, Frank Greensword, dealt very much with MAN Turbo AG (manufacturers of turbomachinery.) in Munich and as I had become quite a proficient German speaker it was decided that I should brush up my French (?) for my future involvement with French companies. One visit I made to France with Gerald Davis a project development engineer, was to Safran Aircraft Engines, outside Paris, to discuss the test programme for the proposed RB207 engine, then a possibility for the 'Airbus'. The project did not come to fruition; probably a good thing in view of the massive technical problems which occurred on the RB211 engine later."

The Rt Hon. the Lord Tebbit CH, PC former RAF pilot and Conservative politician gives an account of his flying experiences with the early Rolls-Royce jet engines.

"In my time as a pilot in the Royal Air Force and Royal Auxiliary Air Force I flew a good many hours powered by Rolls-Royce engines. I flew the Gloster Meteor with Rolls-Royce Derwent Engines, then in the British Overseas Airways Corporation (BOAC) I flew the Avro York and the Canadair C-IV Argonaut, both powered by the Rolls-Royce Merlin piston engine and much later, the Boeing 707-436 with the Rolls-Royce Conway engines.

The Derwent engine, particularly the later versions which powered the Gloster Meteor F8 aircraft and later marks, was a rugged and reliable engine. On one low-level flight at 480 knots at 200 feet over East Anglia near the Bedford Levels, I led a formation of four Meteor F8 aircraft into a flock of seagulls. Every aircraft received some damage. My own had a seagull which penetrated the wing leading edge spar, coming to rest against the main spar, several others spread over the nose cone and rear fin and the remains of a couple more were found on the front impeller of the starboard Derwent engine, which scarcely 'coughed'.

The Merlin engine which powered the Spitfire and Hurricane in the Battles of France and Britain, as well as the Defiant and Mosquito fighters, also powered the Lancaster heavy bomber and the York transport aircraft. I flew a few hundred hours on the York, but I found it liable to serious overheating at low speeds as on take-off and climb power in the tropics. In those conditions it was somewhat unreliable. I also experienced shutdowns due to bird ingestion in flight."

ORGANISATION AND OPERATING PHILOSOPHY OF ROLLS-ROYCE LIMITED 1950-1971

In order to better grasp some understanding and insight of the organisational philosophy and company ethos within Rolls-Royce in the 1960s, it is interesting to share the approach and ethos on organisation, with the emphasis being on development and welfare offered to its employees which Sir Denning Pearson, K.T., J.P., CEO and Deputy Chairman of Rolls-Royce, outlined in a paper written and presented to the London School of Economics and Political Science on November 17th, 1964. This was progressive and forward thinking for its time, particularly with its approach to employees. By his own account, Sir Denning was a 'disciple' of a management philosophy espoused by Arnold Toynbee.

The approach outlined below is taken from selected extracts from this 1964 paper.

1. INTRODUCTION

"... Rolls-Royce is a public company incorporated in England under the Companies Act. Our authorised capital is £30M of which £25,216,722 is issued. The shares are widely held by the public and by the institutions[1] *and there are no individual large holdings. The stockholders, number 45,000. The company manufactures aero engines, motor cars, petrol engines, diesel engines, multi-fuel engines, and industrial gas turbines, railway traction units, shunting locomotives, rocket motors and naval nuclear propulsion units.*

The largest part of the business is in the manufacture of aero engines and of over 43,000 employees some 34,000 are engaged in aero engine work ...

1 By way of example GREA – Guardian Royal Exchange Assurance were the main debenture holders at the time

2. PEOPLE

Organisation is merely a framework designed to enable people to work effectively together towards a common end, and it has always been one of our principles at Rolls-Royce to try and put people first.

Broadly our philosophy has always been to try to create the conditions in all sections (of) the company which would give all our employees the opportunity to develop their talents and capabilities to the full. At the same time, we do all we can to treat everyone employed by the company as a unique individual.

We start by trying to recruit the highest grade of individuals we can at all levels, starting with trade apprentices, who we recruit at 15 and until recently taught in our schools outside the factory until they reach the age of 16 ...

We follow on by recruiting boys who leave school between 16 and 18 into drawing offices and technical apprenticeships, and boys who leave at 18 into engineering apprenticeships

... Finally, we recruit university graduates for a further two-year training scheme. We endeavour to recruit these people in all ranges of our activity, although naturally the largest numbers are engineers and scientists. We deliberately recruit a number of arts graduates.

We now accept that having recruited them and given them initial training we shall probably have to continue to give most of them some continuing form of training for a great part of the rest of their life with the company.

... The following are a few of the specific arrangements which have evolved and ... arisen out of our general attitude towards the people we employ.

I – Workers Shares

In 1920, Rolls-Royce became one of the first organisations to introduce Workers Shares. At that time the company was pioneering in the field of industrial relations.

... it has always ... operated on the fundamental basis that one-tenth of the company's authorised capital is in workers shares. The holders have the same voting rights and interest in the equity as ordinary shareholders.

Workers shares are sold to employees at their par value and can only be sold back to the company. They carry the same percentage interest as ordinary shares.

All employees of Rolls-Royce and of subsidiary companies with five years or more of continuous service are entitled to hold stock under this scheme ... At the present time over 12,000 employees hold workers shares to the value of £1,991,722-12s-0d

II – The Long Service Award Scheme

"I have always held the contention that we have special responsibilities to our many employees who have devoted a large proportion of their working lives to the service of the company.

A positive attempt to discharge this responsibility was made in 1959 with the introduction of the Long Service Award Scheme. Under this scheme financial benefits are extended to all employees with more than 25 years' service in the event of sickness. Provision is made for a scale of continuing payments to those suffering from a long-term illness or who become permanently or totally disabled, right up to retirement age.

III – The Works Contract Scheme

Contracts of employment with individual employees date back a long way in the company and first started in our Experimental Department between the wars. Some ten years ago, however, our Works Contract Scheme was introduced as a new concept in labour relations. Originally brought in for skilled workers, and recently extended to cover semi-skilled grades as well, the scheme aims to increase both the dignity and financial security of our best manual workers. . .. Individuals are chosen for their industriousness and conscientious service by foreman and supervisors. Under the scheme contract holders receive a bonus payment on paid holidays and a wage for holidays which are normally unpaid. In addition, payments are paid during periods of illness and for those who subscribe to the company Pension Fund there are settlements on death before retirement age or for permanent disability. The contract requires that three months notice should be given on either side of termination of employment.

3. *ORGANISATION*

Broad structure

Up until the early 1930s, the Board of Rolls-Royce was remote ... in London and on the rare occasions on which it visited Derby, met in an hotel rather than at the works. The headquarters are now firmly established at Derby where the Board normally meets and eight of the twelve directors are executives, seven of whom live in the Derby area.

The company is wedded to the concept of a mixed Board of full-time executive and outside directors.

Objectives of the Company

I) I believe that in essence any industrial activity is basically directed towards providing a service through meeting a need. In our case, the need which we meet is for a large spectrum of power units in the gas turbine, piston engine and nuclear fields and motor cars of the highest quality. In meeting this need we aim to maintain our existing customers and to obtain an increasing share of the world market in these fields. We also endeavour to fulfil any other needs which arise throughout the world, which can be met by development of the knowledge, experience and technology that we have accumulated through existing activities. This is a very broad and fundamental objective and one which in my opinion appeals to our employees and is acceptable to our shareholders.

II) In the long run the objective of earning the maximum possible profit for our shareholders on the capital they have entrusted to us can, in my view, only be fulfilled in this way.

III) A further objective is to fulfil our obligations to our employees. The most important of these is, I believe, to take all possible steps to ensure security of employment. In this connection we endeavor to follow the policy of sub-contracting the peaks of our working load over as wide a field as possible.

IV) We have a responsibility to our suppliers who also have employees and shareholders. An objective is to give them reasonable warning of changes of programmes and to trade fairly with them.

V) We all have responsibilities to the communities in which our factories are situated and, on a broader basis, to the nation.

Philosophy of Management Organisation

The problem of organisation, which is that of coordinating the activities of a large number of people to achieve a given but changing objective, is in effect one of making the maximum possible use of their total abilities and in particular their total capacity for creative thinking and for decision making. This leads to the general principle of trying to organise in a way which pushes down into the organisation as far as is practicable all responsibilities for decision making.

I believe that the degree of formalisation in a company organisation is very dependent on its stage of growth. In periods of rapid expansion, the maximum possible fluidity must be permitted so that the changing natures of the majority of both internal and external factors involved can be met. Once the growth has stabilized and the external needs have been fully assimilated then a much more clearly defined structure has its merits.

We have for some considerable time subscribed strongly to the modern conception of separate operating divisions and a central functional control.

Operational Control

We believe that our operational activities are best broken down into the maximum number of divisions or parts which can constitute satisfactory self-contained working units. In general, I define a self-contained operational activity as one that is capable of being made responsible and accountable for the profitable use of definable assets. For manufacturing activities this implies the capability of sustaining engineering and selling departments, and of making a trading profit. Each unit has to be large enough in relation to the type of products to provide a manufacturing organisation of the optimum economic size. In practice, the units are divisions or subsidiary companies under the control of managing directors or general managers.

Functional Control

Our headquarters organisation consists of a number of functional executives who are responsible for covering the main functional activities of the company. In addition to their purely functional duties, some also carry responsibility for the operations of specified subsidiary companies in the United Kingdom which are not the direct responsibility of the managing directors or general manager of the main operating divisions and are responsible for watching the company's investments in associated companies. . . .

Implementation

Up to the second World War, Rolls-Royce was a comparatively easy company to manage. It was small by present-day standards and its products confined to motor cars and piston aero engines which were designed and built side-by-side and of a comparable technical standard.

The events which have taken place in the years since 1945 have made it necessary to develop new techniques in management and a new structure.

When the war ended, we made the decision to recommence manufacture of the motor car at Crewe. This led naturally to the separation of our two main lines of activity- aero engines at Derby and Glasgow and the motor car at Crewe.

As the markets and customers for these products were so different, each had separate sales and service organisations built up to serve them. It was equally natural that each should maintain its own engineering and development organisations.

In 1954, a formal distinction was made between the two organisations with the introduction of the appellation Aero Engine Division and Motor Car Division. As our oil engine business developed it seemed only natural again for it to become a self-contained unit and the Oil Engine Division was created in 1956.

As a development of this system, divisional boards were created in the summer of 1958 and the managing directors of the divisions were vested with full responsibility for day-to-day operations.

From 1958 onwards, the basic principle of running what in fact can to a large measure be regarded as separate companies had gradually been elaborated. We now have a logical company organisation with headquarters exercising functional control over a series of divisions and subsidiary companies.

Headquarters are responsible for long range planning of the company and investigation of new business opportunities in the broadest sense. I would like to lay strong emphasis, however, on the fact that our basic method of extending spheres of activity is through the divisions. Thus, our nuclear interests started from the possibility that aero engines might one day run on nuclear fuels. As described previously, Rolls-Royce and Associates grew from this work. Again, the realisation that aero gas turbine engines could serve in industrial applications came from the Aero Engine Division.

4. CONCLUSIONS

I would summarise what I have learnt about organisation and Rolls-Royce specifically, as Chief Executive of the company for the last seven years, as follows:

1. The requirements of people and the maintenance of an environment in which their creativity is stimulated to the maximum is very much more important than the organisation, although their efforts can only reach fruition through the organisation.

2. In order that the best use may be made of creativity, the responsibility for making decisions must be pushed well down into the organisation.

3. At the present stage of the developments of management techniques, I consider that a divisional form of organisation of the type which I have described earlier in the paper is best suited for any company which serves a number of differing markets with different products.

4. There is not and never will be, any finality about the best form of organisation. If it were possible to create an ideal organisation to deal with a given set of external circumstances, the very success of that organisation in dealing with the external circumstances would create a new situation . . . In this I am a disciple of Toynbee. In my view and experience it is essential to keep the organisational pattern fluid to make certain it is never allowed to set and to be constantly taking a fresh look at it in light of today's circumstances.

5. That industrial management reached the stage some time ago when the accumulated experience obtained in practice was capable of being codified and thus passed on in the form of formal education like any other discipline.

This does not mean that a degree in management studies will produce a first-class manager any more than a degree in music necessarily produces a first-class musician. It does mean, however, that those already carrying responsibility for managing other people must be encouraged to accept the advice of those trained in management techniques, just as they are today prepared to accept advice from lawyers on matters of law and doctors on matters of health.

> *6. It is most important to avoid the temptation of becoming interested in the organisation that one forgets that it is merely a means of managing the company efficiently and not an end in itself."*
>
> **Sir Denning Pearson, K.T., (1963); J.P. (1964);**
> **Deputy Chairman and Chief Executive, Rolls-Royce Ltd. 1957-1968;**
> **becoming Chairman and Chief Executive 1969-1970.**

ROLLS-ROYCE LIMITED AND ITS MAJOR UK DIVISIONS – PRE-RECEIVERSHIP

The divisions operated like a hub and spoke network centering on Derby and the Derby Engine Division as its main base and headquarters.

Following the takeover of Bristol Siddeley Engines by Rolls-Royce in 1966, these subdivisions were eventually rationalised and reorganised as set out below.

Derby Engine Division (DED), Derby
Bristol Engine Division (BED), Bristol
Small Engine Division (SED), Leavesden
Industrial Marine Division (IMD), Ansty
Motor Car Division (MCD), Crewe
Oil Engine Division (OED), Shrewsbury

In 1969, the Industrial Reorganisation Corporation (IRC) had recommended amalgamating SED and the BED and selling others including the OED. However, these remained until the receivership period when both the Motor Car Division (MCD) and the Oil Engine Division (OED) were subsequently separated out and listed on the stock market.

In 1971, Rolls-Royce Limited was the only major British aero engine manufacturer. At its headquarters in Derby, the Derby Engine Division (DED, formerly the Aero Engine Division) had a workforce of approximately 20,000, whose prime activity was aero engine manufacture.

The company also employed approximately another 60,000 people worldwide including the other five core subdivisions around the country. Subsidiaries included, for example, Rolls-Royce Canada, and associated companies such as Rolls-Royce and Associates Ltd.

Rolls-Royce in Scotland

Rolls-Royce's presence in Scotland began in 1939 when in support of the war efforts at the start of World War II, Rolls-Royce opened up a "shadow factory" in Hillington, Scotland to provide additional production capacity for manufacture of the Merlin Engine for the RAF fighter and bomber planes.

Later, the East Kilbride (EK) factory opened in 1953 and along with Hillington became part of part of the Derby Engine Division (DED) group. Around 1962 following a McKinsey (Management Consultants) report EK became known as the Scottish Area Overhaul Centre, Works Centre 9 and Hillington became the Scottish Machining Work Centre purely concerned with component manufacture.

In 1968 the Rolls-Royce Scottish factories covered 5,449,970 square feet, known as DED Scottish Factories, and consisted of Hillington, East Kilbride, Hamilton and Blantyre and a number of small subsidiary sites in and around EK, Hillington and Glasgow and post 1971 were generally referred to as DED -Scottish Group. (Both the Hillington and EK factories closed in 2005 and 2015 respectively).

Rolls-Royce in Derby (DED)

The company's persona and outward image remained very paternalistic, where whole families of people were employed for large parts of their working lives, and the expression 'a job for life' was certainly the case for many who worked there. The days of a future diverse, global and multicultural workforce being an everyday part of the company's culture was still many years away. In fact, most of the employees in Derby were born or emanated from within the immediate Derby/Derbyshire area and often generations of families would work there.

The management and leadership style at Rolls-Royce at that time could be fairly described as slightly 'old school' with rigid chains of command across senior management, despite Sir Denning Pearson's ethos and the implementation of many of his employee initiatives. The majority of employees had spent most of their working lives at Rolls-Royce since the period immediately following the end of World War II.

Before the receivership, most of the works sites in Derby were spread out across numerous urban and suburban areas, such as Main Works and the Sinfin sites in Derby. Many of these sites have since been vacated (including Gresham Road, Osmaston Road, Mickleover Station Road, Main Centre, Colyear Street, Sandown Road and the Old Hall, Littleover) either as part of the receiver's realisation of assets or subsequent rationalisation. Additional sites included Mountsorrel in Leicestershire and Hucknall in Nottinghamshire. The primary engineering and commercial sites remain much the same today, i.e., Moor Lane, Sinfin A, C and D sites, B site having been demolished in Spring 2023 and Raynesway (Rolls-Royce Nuclear). Manufacturing has moved from Main Works to Sinfin where a new Precision Casting Facility (PCF) and Precision Testing Facility (PTF) have been built. Development and production testing are now carried out in new test beds on Sinfin D site.

ROLLS-ROYCE CORPORATE GOVERNANCE – PRE-RECEIVERSHIP

The Boards The governance and operating structure of Rolls-Royce had changed little since the 1950s. It comprised a main board of directors and also divisional boards of directors. Members of the divisional boards controlled the day-to-day running of the subdivisions as well as a network of subsidiary and associated companies located in the United Kingdom and overseas. Each of the divisional board's managing directors was accountable to and had a seat on the company's main board. Many of the main board members, 'career engineers' by profession, had worked for Rolls-Royce all their working lives.

The Main Board would normally meet six times a year in alternate months with additional meetings to consider the annual accounts and half yearly financial statements. The board delegated its authority to the Chief Executive, a post held by Sir Denning Pearson from 1957 to November 1970. Consideration of the annual accounts and the half yearly financial statements would be on the agenda and together with past and future performance, would be presented to the board by the relevant divisional managing director or executive director of the board.

Directors on the board with divisional responsibilities were available for questioning on divisional activities but each representative would not have much knowledge of the affairs of the other divisions. Financial reporting would be provided by way of information within the board papers in the first section of a red folder, the second section would include more of a narrative report by the divisional directors on what they considered should be reported and this was also dependent on what had been the outcome of the preceding Chief Executive's meeting. The red folder could contain some 70 pages of information. A green folder would contain the agenda for the meeting and after 1969, with the formation of the main board finance committee, the minutes from that. The Agenda would later call on board members to 'note' the items in the red folder, which included financial information. The green folder would also contain further reports from the chief executive, the group executive directors and the divisional managing directors setting out any matters which needed a board decision. As detailed in the chronology of background events later in Chapter 6, some main board directors stated they were not made aware of the 'Metcalfe Report' submitted to the DED board, which indicated in July 1969, the unsustainable situation of the RB211 development programme costs and reported an 'adverse variance', a loss, against a previously reported profit and did not seem to have been reported in the main board minutes as having been mentioned.

Divisional boards were created in July1958 with an amendment to the company's articles of association for the Aero-Engine, Car, Oil Engine and International divisions for the management and carrying on the business of those divisions; (the international division was disbanded in 1964). Divisional boards were responsible via the divisional director to the main board.

The Derby Engine Divisional board would also meet six times per year in advance of the main board meetings. It had no legal significance with regard to the Companies Act 1948-1967.

The DED board included representatives of five separate departments: engineering, experimental, production, commercial and financial. However, this was considered by some as more of a forum for management communication and advisory functions rather than taking decisions by majority vote as with a main board. Sir David Huddie had final say on the day-to-day activities of the division as full responsibility for it was carried by him. He would hold regular Monday morning meetings when he was not in the US. In 1969, there were 22 members of the DED board, all of them men and only one who was not an engineer.

The Chief Executive's Meetings: These would be held soon after a divisional board meeting and before the main board meetings, with representatives from each division. One of the purposes of this, the Government Inspectors MacCrindle and Godfrey were later told, was to consider what information from the divisional board meetings should be placed before the main board. R.A. MacCrindle, QC & P. Godfrey, FCA, *Rolls-Royce Limited–Investigation under Section165(a)(i)of the Companies Act 1948, Her Majesty's Stationary Office, 1973, (pg 40).*

Committees

The main board set up sub-committees for specific purposes with delegated authorities, the two main ones being the Finance committee established in 1969 by Sir Denning Pearson and the Executive committee set up in November 1970. The finance committee was composed of Sir Denning; deputy chairman Mr.

Whitney Straight; the company financial controller Mr. Trevan-Hawke; the company commercial director Mr. Hinkley plus two non-executive directors, Messrs. Gill and Smith. It was not until May 1970 that a divisional finance committee was formed.

The executive committee was formed to take over the responsibility from the chief executive, with Sir Denning resigning the same day that committee was set up.

In February 1967, following a presentation to the main board sub-committee which, showing a general decline in sales of the current engine portfolio from 1969 through to 1975, a decision was taken that the RB207 should be offered to European Airbus and the RB211 to both Lockheed and McDonnell Douglas respectively. Without the development of a new engine Rolls-Royce would not be a competitor in the big engine market. This recommendation was then formally approved by the main board in July 1967.

1. Rolls-Royce Silver Ghost production line at Nightingale Road, circa 1909.

2. Sir Henry Royce at the wheel.

3. Early Rolls-Royce car built in 1907 – The Dreadnought Chauffeur Training Vehicle.

4. Rolls-Royce's iconic Silver Ghost.

5. Rolls-Royce Cavalcade – A young Raymond Baxter, as passenger, former BBC presenter at Farnborough Airshow

6. Aerial View of Derby – Main Works, circa 1960s.

7. Aerial View of Rolls-Royce Moor Lane site, Engineering Block, circa 1960s.

8. Exterior wall, Rolls-Royce, Castle Donington site, next to the raceway.

9. Main Works, Nightingale Road Entrance, circa 1950s.
The original company HQ block until the late 1960s.

10. Moor Lane engineering block, circa 1968.

11. Rolls-Royce Alfreton Road site, Derby 1967.

12. Rolls-Royce building, Marble Hall, Nightingale Road –
The Battle of Britain memorial window.

13. Rolls-Royce Castle Donington, drop forge in action.

14. Rolls-Royce Eagle Engine production line, Nightingale Road, Derby.

15. Rolls-Royce site, The Old Hall, Littleover, Advanced Research Department HQ, circa 1960s.

16. Rolls-Royce Derby sites – Slack Lane and Gresham Road.

17. Senior Rolls-Royce Executives and French Representatives (Airbus) outside the Engineering Block, Moor Lane – RB 207 Meeting, circa mid 1967

18. Rolls-Royce Dinner Function, R – Adrian Lombard, L – Ernest Eltis.
2nd left David Kaye with Representatives from Airbus in attendance circa mid 1967

COPYRIGHT ACKNOWLEDGEMENTS

Images 1–12 & 14–18 copyright Rolls-Royce Limited, with kind permission of the Rolls-Royce Heritage Trust.

Image 13 by courtesy of Richard Smith.

CHAPTER 2

THE CIVIL AVIATION MARKET; THE RB211 ENGINE OPPORTUNITY AND STRATEGY

"Building the RB211 would not guarantee we stayed in business. Not building it would guarantee we went out of business."

SIR DENNING PEARSON (1967)

THE NEW AGE OF LONG-HAUL CIVIL AIRLINERS

From the late 1950s, the advent of long-haul travel for passengers became a reality with the introduction of the Boeing 707 and the Douglas DC series aircraft, some powered by Rolls-Royce Conway engines, into passenger-carrying service with their ability to cover non-stop long-distance transatlantic flights.

In parallel with the reshaping of the British aircraft industry in the 1960's, it was apparent that Rolls-Royce needed to look to the future of its engines portfolio given its declining income and profit forecasts by the end of the decade, due to both the reduction in military business and the declining sales forecasts for the older civil engines. The portfolio of both small and medium, civil and military engines included the Dart, Spey, Avon, Conway and Tyne. As an example of the engines that Rolls-Royce was able to apply to both civil and military applications, the Avon engine was installed in the Comet 4 airliner, and military variants in the Hawker Hunter and the English Electric Lightning aircraft.

The company believed that its very survival and future prospects would lie in the next generation of large engines and the market that was predominantly in the USA would become a major target for its marketing efforts.

In competition with Pratt & Whitney engines, Rolls-Royce introduced the two shaft Conway turbofan engine in 1960, which was largely designed and inspired by Adrian Lombard, a much-respected Rolls-Royce engineer. It was the world's first turbofan engine to enter service. The design features of the Conway engine showed this was a new bypass engine, significantly different from that Pratt & Whitney was offering, with the JT3D which also powered both the Boeing 707 and Douglas DC 8 aircraft.

ROLLS-ROYCE AND HMG FUNDING FOR LAUNCH COSTS

The objective of Launch aid was to limit the amount given out of the public finances. It was an interest free, financial contribution by government toward the costs of design, development, production jigs and tools and the higher labour costs involved in the early production stages of either civil aircraft or aero engine projects. The aid was, however, repayable in the form of a levy, payable to the government, based on a share of the forecasted margin between sales and the estimated manufacturing costs on subsequent engine sales and spare parts. In 1960, the Conservative government policy for direct financial assistance on launch aid for civil aircraft would be reintroduced and the government's levy on sales would be more formally applied than in the 1940's.

The government limit on launch costs reviewed in 1960 would, for engine development, be no more that 50% and if the amount required, such as on development or manufacturing costs, detrimentally affected the manufacturer's margin, it would still be expected to pay the levy to the government and take the reduction of profit 'on the chin'.

The idea was that the risk and possibility of profit resided with the company, to provide 'a commercial incentive and spur to exercise financial judgment'. R. A. MacCrindle, QC & P. Godfrey, FCA, *Rolls-Royce Limited, (1973) (pg.11)*

The future policy toward the industry in support for civil projects had therefore been set.

During the late 1950's through to the 1970's the subject of access to government funding assistance in the guise of launch aid for new engine projects became a continual issue for Rolls-Royce. Sir Denning Pearson was not slow in lobbying the government on several occasions about this, particularly with the advent of the 3-shaft civil engine project, being developed by Adrian Lombard, on the horizon. Referenced in the above Report Ibid (*pg.12*), that a meeting had taken place in March 1958, between the major manufacturers and the then Ministry of Supply[2], announcing amongst other matters, changes in the way the Government would only award specific contracts for work it actually required, rather than providing out and out investment for general research by way of annual awards. When it was pointed out to the Minister that those types of projects provided technological advances for applications other than military his response was that it was 'not the Ministry's intention to discourage firms from financing their own level of research projects above those the Ministry could afford to support'. (See Joe Clayton's remarks in Chapter 5).

2 Ministry of Supply (MoS) was a government department set up in 1939 engaged in coordinating supplies of equipment to the armed forces. This absorbed the separate Ministry of Aircraft which was responsible for aircraft production when that was abolished in1949. The MoS was, itself, abolished in 1959 and replaced in part by the Ministry of Aviation.

Rolls-Royce representatives argued that if it were to survive it would need to feel it had the full backing of HMG. The Minister informed them that HMG stood firmly behind the industry and would support it 'to the full extent of its power'.

The following is an extract from a paper with reference to launch costs, submitted by Sir Denning Pearson, Rolls-Royce Chief Executive to the Ministry of Aviation, July, 1961 and demonstrates the high cost of research and development, (R&D), the timescales involved in development and that profitability comes later, mainly from the volume of spare parts sales and services.

> ..."5. *Our best estimates show that neither in volume nor profit will the turnover of the years ahead be sufficient to carry present launching costs while maintaining the ability to undertake profitable new projects as they arise. The point has therefore been reached when the company will be prevented from achieving the fullest exploitation of the position which has been won in the world aero engine business without further government support.*
>
> 6. *No project is started without the most careful scrutiny of its commercial prospects and possible profitability; but no matter how successful any one project may be, the time which elapses between the commencement of expenditure and the delivery of a sufficient quantity of engines and spare parts to recover the launching costs may be as much as 15 years. The length of this period is a reflection of the highly competitive prices which limit the rate at which launching costs can be recovered from sales revenue.*
>
> *This cycle of expenditure and recovery means that large sums of money are unproductive for a considerable time and that the total amount unrecovered will clearly increase as new projects are started and will only reach a peak when the revenue being derived from current projects is enough to provide development funds for new projects. We confidently look forward to this situation, but it would be optimistic to expect to see this happening for several years bearing in mind that is only some six years since the first major private venture project was started."* ...

It is acknowledged that profits from sales of large volumes of spare parts provide a greater profit margin than on engine sales and that this of course does not happen until some years after the engine is installed. The above comment highlights that profits from sales of spares, when eventually received, would provide investment for new projects.

The expectation of government support was again referred to in a paper presented by Lord Kindersley, then Rolls Royce chairman in July 1962 to the Ministry of Aviation which highlighted the financial position of Rolls-Royce at that time with regard to the amount of money it had contributed to development of civil engines in the years prior, whilst waiting for a Government ordered report to be produced and recommendations implemented regarding the financial support it could expect. This was referred to as the 'interregnum' period. Rolls-Royce had expended a lot of its own money on the R&D relating to the Tyne, and later the Spey & Conway engines, Rolls-Royce continued its R&D funding of the Tyne and later the Spey and Conway engines said by Lord Kindersley, then Chairman, in order that 'by this action . . . engines (have) been available for the British airline aircraft programme'. Basically, airframe manufacturers were receiving development funding as opposed to engine manufacturers who had continued to develop engines in the meantime, self-funded.

It was well aware of the potential problems in developing a civil engine where the maximum support on an estimated amount for R&D would be 50%. It pointed out the considerable cost already expended by the company along with its ability to capture the international market and particularly stating it led the world in development of the 'bypass engine'.

The US aircraft industries on the other hand were able to develop their R&D fully supported by the Government with most of this being undertaken on military projects which were then applied to civil engine development, giving the US a distinct advantage.

The paper ended on the following note which would resonate down the years to the RB211:

> **"We have been told on more than one occasion that special consideration would be given if the future of our business was in jeopardy because of lack of finance".**

In Lord Kindersley's opinion that point had been reached.

Sir Denning Pearson, in February 1963 also outlined in a letter to Sir Arthur Vere Harvey who was Chairman of both the aviation and defence committees, in the strongest terms of the lack of investment for civil engine development. He set out that the Government contribution for the Tyne had been £4 million out of at least £16 million spent and the Spey £3 million out of £20 million spent. This in effect was less than the 50% commitment in the policy that had been announced in 1960 by the Conservative government. Financially, proceeding with the development of the Spey was said by Sir Denning to have brought the company to, very near, financial breaking point. In recognising this, the Ministry of Aviation tried to obtain a further £1 million for Rolls-Royce toward the Spey, but they were refused by the then Chancellor of the Exchequer.

This lobbying of the Government was not the first nor would it be the last time Rolls-Royce would seek support for its R&D to secure successful project outcomes.

THE UK POLITICAL LANDSCAPE AND THE RE-ORGANISATION OF ITS AEROSPACE INDUSTRY

The decade of the 1960's swept in a period of great social change and opportunity for the United Kingdom both on the political and economic fronts which had profound effects for both the country and Rolls-Royce as a company.

The general election of 1964 marked a change in politics from a Conservative government which had been in office for the previous 13 years, to a Labour government which would stay in power until the end of the decade. Harold Wilson's government rode an initial wave of optimism for Britain (which had dissipated by the end of its term of office). The Labour government inherited a very weak economy followed by a period of deflation, public expenditure cuts and, eventually, the devaluation of the pound.

It took office with a manifesto called 'Time for Decision' which promised public ownership of the ports, creation of a National Freight Authority and public participation in the aircraft industry. The incoming Labour government had focused on new technologies in its manifesto, and it aimed to use science and technology to reinvigorate British Industry and inject a boost to the economy. Roy Jenkins, appointed as Minister for Aviation, viewed that the industry had been supported with favourable economic conditions for too long. Lord Plowden was to be commissioned to conduct a review of the aircraft industry.

> *"The Labour Government elected in 1964, came to power determined on an expansion and redistribution of national resources allocated to technology. In opposition, the labour party had seized on technology as an electoral issue, criticising both the emphasis and implementation of Conservative policy. According to Labour, science and technology, well used could regenerate the British industry and revitalise the economy. The Conservatives had wasted the opportunity by concentrating on defence and prestige projects. The Labour Government, on the other hand was determined to 'back winners', commercially viable projects*

and areas of technological innovation which could benefit a wide range of industrial applications. Consequently, it embarked on an ambitious and unrepentantly interventionist strategy to provide British industry with the conditions to advance and exploit technology. At home Governmental reorganisation (the creation of a Ministry of Technology), the encouragement of mergers (to increase the capital and productive base for R&D investment) and the identification of 'National champions' (for example, BLMC in cars, ICL[3] *in computers and Rolls-Royce in aero engines around which to base public investment. Internationally, the Labour Government responded to the 'technology gap' syndrome by asserting even more than its predecessor the need for European collaboration. This still had a practical purpose, to share costs and widen markets; but like its predecessors, labour accepted that technology was also politically useful in demonstrating the United Kingdom's value to the European Economic Community. In short, technology was again a strong bargaining card in negotiations to join the Common Market, or as a link if Britain were refused entry".*

Professor Keith Hayward FRAeS
(1983) *Government and British civil aerospace.* A case study in post-war technology policy; Subsonic Politics, 1964-1970, (pgs. 69-70) Manchester University Press

THE SIXTIES – An account of life in the company pre receivership by Sue Allen, former Rolls-Royce engineer.

"Life in Rolls-Royce immediately prior to 1971 was a very different time from now in the twenty-first century and possibly can explain why 1971 had such a huge impact on the workforce and the towns and cities which depended on the company's survival.

This was the time of the standard long working day (42 hours/week in the mid-1960s) with the luxury of an hour at lunchtime. In addition to serving lunches, the Moor Lane canteen also provided 30-minute teas for those who wanted to work on after 5.30. Managers had their own dining room with waitress service and china cups!

A large proportion of people smoked, not just cigarettes but cigars and pipes. How any of us survived the fug-filled meeting rooms with ashtrays complete with pyramids of nub ends, to say nothing of the accidental (usually) fires in wastepaper baskets, I really don't know!

3 BLMC: British Leyland Motor Corporation Ltd.; ICL, International Computers Ltd

There were surgeries on most sites where employees could go if they felt ill (after breathing in all that smoke probably); families were not forgotten with Christmas parties complete with presents provided for children. Free Christmas turkeys were still given to employees until the early 1960s! It was unusual to meet a manager or supervisor who hadn't risen through the ranks, often with the people that he had known at school. Brothers worked alongside brothers and cousins with cousins. Whether this pride in old-established working practices at the expense of new ideas was a good or a bad thing is open to argument. It was often regarded as the 'job for life' and Rolls-Royce employees were fiercely loyal.

The everyday experience of being a woman in engineering was very different then.

In the 1960s, I had decided to study engineering at Derby Tech (now the university) by part-time day release which was the norm for many male apprentices but certainly not for women. It was just the persistence of my manager that made the company, reluctantly, let me study. "Well, I suppose we could give it a go," was their response. The proviso was that, although the men continued to have their fees paid for them at the beginning of each academic year, I had to pay my own and could only claim them back at the end of each year if I passed my exams. I did.

It appears to have worked out to everyone's benefit as, after the redundancy I returned in 1974, retired in 2004, worked for another 15 years as a contractor and am now involved with the Heritage Trust. I hope that the company view its investment in my education as a profitable one.

Women did, however, have some perks in that we were allowed extra time off in the middle and at the end of each day. This totalled 1 hour 15 minutes per week and was announced over the Tannoy as "Female staff employees may now leave their work". The reasons given for this ranged from being able to powder our noses properly, to enable us to get a seat on the bus or, in those politically incorrect times, to let us get home so that we could start our husband's tea. These few extra minutes only partially justified being paid 20% less for the same work with separate pay scales for men and women. The maths didn't add up then either, but it was best not to argue.

On one occasion I was sent home by the Moor Lane commissionaire for arriving in trousers. Two buses each way home and back and I returned in a dress.

In the late '60s and early '70s we could say that we lived in interesting times!"

The Three Shaft Engine; High Bypass Pressure Ratio (HBPR) The Foundations of the RB211

In 1963, Adrian Lombard began laying down the foundations for the design of a completely new large aero engine, capable of competing for the new generation of long-distance aircraft then entering the market, i.e., the Boeing 747, the Douglas DC-10 and the Lockheed TriStar. It was one with the capability of thrust required for these types of aircraft.

Rolls-Royce now had its own large aero engine design on the drawing board and had started work on the demonstrator engine contract awarded by Her Majesty's Government for the RB178 engine, funding a contribution of £1.3 million in total.

"Evidently, a new, large engine would be required to replace the Conway although there was minimal activity at this time by the aircraft manufacturers. The USA was focused on a new Supersonic Transport Aircraft for commercial applications and the Heavy Lift Military Transport aircraft which became the Lockheed C-5A Galaxy. Pratt & Whitney and General Electric were also running engine demonstrator programmes in support of the Heavy Lift Transport requirement. In Europe, Vickers, which had now become part of the British Aircraft Corporation (BAC), was studying growth versions of the VC10 whilst De Havilland, which was now part of Hawker Siddeley, was working with the European aircraft manufacturers on a new aircraft which was to become the Airbus A300.

Rolls-Royce designed the RB178-14 for the VC10 and European Airbus applications, a two-shaft engine with a bypass ratio of 2.0 which ran as a demonstrator engine in June 1966.

Phil Ruffles CBE, FRS, FREng. (2020)
(Former Rolls-Royce Director of Engineering and Technology)

During 1963, Rolls-Royce had organised a showcase conference to display the quality and range of the company's work, particularly the research and development work being carried out at the Advanced Research department based at The Old Hall in Derby. This conference would eventually turn out to be the catalyst that Adrian Lombard needed to make the breakthrough to a new range of high powered 3-shaft aero engines based on four main characteristics.

Firstly, whereas previously high-power engines had been based upon two concentric shafts, with one rotating inside the other, Lombard had initiated design work on making the next big step to three concentric shafts. That in itself was a very bold move which no other designer and manufacturer of aero engines had yet attempted. It was an arrangement which was potentially full of technical problems, but the advantages in engine performance justified the attempt to solve them. Lombard had already set up a team of design engineers to unravel the difficulties under a brilliant engineer named Geoff Wilde who took the design forward.

Next, he wanted to incorporate the latest developments in the cooling of turbine blades. He was also determined to design every conceivable source of unnecessary noise out of the engine so that these new engines would be the quietest ever. Finally, he very much wanted to be able to use new carbon fibre reinforced materials in the new design of engine; Hyfil, which was manufactured entirely by Rolls-Royce at the Hucknall site, from acrylic fibre tows using a process that had been developed jointly with the Royal Aircraft Establishment (RAE) with support from academia.

Gradually we were seeing the birth of a revolutionary design of a three-shaft aero engine which was later to become known as the RB211, ahead of its time, and more technically advanced than its competitors, a game changer in every respect.

Phil Ruffles again:

> *'Whilst three-shaft engine studies started in 1963, Geoffrey Wilde, who was the engine's champion, wrote in 1964 that:*
>
> *"The three-shaft engine was the optimum configuration for bypass ratios of 2.5 and above and an overall pressure ratio greater than 22:1."*
>
> *It was becoming apparent by the mid-1960s that large engines of high efficiency HBPR (High Bypass Pressure Ratio) were the future and would capture a significant portion of the civil aero engine market in the 1970s and beyond. In 1965, when the Lockheed C-5A was launched, powered by the TF39 engine with a bypass ratio of 8.0, Pan American Airlines asked Boeing to design an aircraft with 2.5 times the seat capacity of the B707. This aircraft became the B747. Rolls-Royce offered the RB178-51 with bypass ratio of 6.0 (higher than the Pratt & Whitney offering) for the B747 but the contract was awarded to Pratt & Whitney with the JT9D which had benefitted from their demonstrator engine programme. Thus, the first wide-body airplane powered by a high bypass ratio engine was born.*

Following the launch of the B747, in April 1966 American Airlines put out a requirement for a transcontinental aircraft to which Lockheed and Douglas responded with the result that the RB211-22 aero engine was selected in March 1968 to power the Lockheed L-1011 TriStar aircraft. The DC-10 was powered by the General Electric CF6-6, a derivative of the TF39.'

The Plowden Committee Report on the British Aircraft Industry – Cmnd 2853

One of the first major tasks of the incoming Labour government in 1964 was to open a committee of inquiry into the aircraft industry in the UK which reported on its findings in December 1965. The committee had been charged to review and to recommend what the future role of the Industry should be. The Plowden Committee Report portrayed an industry in difficulty mainly caused by the relatively small volumes of production needed in the UK.

Incredibly, the report deemed that there was no predestined place for an aircraft industry in Britain. The economic justification for the industry receiving more government support than other industries rested on whether it provided particular benefits to the country.

In its general conclusions, the report made several policy recommendations, that in order to deal with the problems confronting the industry and its survival, it should, amongst other matters, consider:

- Wholehearted collaboration with European countries on a comprehensive range of civil and military aircraft projects with the aim of evolving an industry which could fully compete with the USA.
- To concentrate on projects where the development costs were not disproportionate in relation to the market.
- Rationalisation of both airframe and aero engine manufacturers, citing that it didn't believe either military or civil aircraft or aero engines would suffer if the competition disappeared between UK companies.
- A sustained drive to increase exports (with designs which would appeal to a wide range of customers).
- The Government taking a partial financial stake in airframe companies in order that it would be more directly engaged in the industry's affairs while at the same time removing the duplication of control systems which obstructed efficiency.

The Plowden committee believed that the industry would emerge smaller and stronger and make a valuable contribution not only to the UK economy but also that of Europe.

Some of Plowden's recommendations to Government as guidance for future policy are summarised below:

- The level of support which the aircraft industry had been receiving from the government was too high and should be reduced; with the aim of creating conditions for a thriving industry with no more support or protection than was given to comparable UK industries[4]
- That the objective should be to improve the relationship between sales costs and development and initial production costs and look to international collaboration as the means of doing so by seeking comprehensive programmes covering a wide range of projects, which was not seen as being achievable with the United States at the time and to focus on association with European Industries within the European aircraft industry[5]
- That where technically and economically sound, discounting future requirements for military aircraft and weapons, British engines and equipment should be incorporated in aircraft bought from the US and its then limited association in aviation with the US could be extended if there was the political will to do so.
- Mergers between airframe groups or between aero engine groups would offer scope for economies in overheads, design facilities and rationalising production.
- The market for civil aircraft was continuing to expand but with a small home market, the short-term prospects for Britain were not favourable. It was however, believed that there were potential future rewards if collaboration with European manufacturers was developed and believed to be the only foundation on which to launch a project to compete with the US.
- The expectation was at that time that aircraft manufacturers should contribute at least 50% of the launch costs of civil ventures. However,

4 Notwithstanding the fact that the final cost of the RB211 could not be assessed, this policy guideline may also have had an influence on the subsequent Conservative 'Heath' Government's reluctance to rescue Rolls-Royce in its entirety, only being interested in those parts considered vital to the National interest and the preservation of Government contracts, being primarily military, alongside of which they had adopted a 'lame ducks' policy of letting private enterprise sink or swim.

5 Airbus was also a future company borne out of this policy.

the government was likely for a time to have to provide more than 50% of the funds in some instances. However, in the case of aero engines, the committee could see no grounds in general for the Government to depart from the then policy of a maximum 50 % contribution toward the launch costs of new projects, although larger grants could be given in exceptional circumstances, but this was not envisaged for engine development. (In the case of the RB211 engine a final launch cost commitment of almost 70% contribution was made). It was also expected, going forward, to play a greater role in the country's civil aircraft programmes in market studies and formulating new requirements.

- Only those projects which satisfied the most stringent tests of 'economic promise 'would be supported by Government funds and manufacturers contributions to launch costs of a project would be a good indication of gauging its commercial merit.

It recommended that partnerships between Britain and France should be sought, being the two largest and strongest aircraft industries outside of the US and given France having several promising projects underway, one of which was the Concorde aircraft.[6]

However, the Report was not altogether well received, not surprisingly at that time, given it openly admitted the country could not, without collaborating with Europe, undertake projects on its own. Furthermore, the Society of British Aircraft Companies thought this would constrain any negotiating position prior to agreements being made. Others thought it did not go far enough, including 'The Economist', which apparently urged out and out nationalisation (the committee had considered this but eventually discounted it). The publication also argued there was a strong case for Britain to concentrate on aero engines, whilst the French, for example, might specialise in airframe production. Notwithstanding the criticism the recommendations were, in principle, accepted by the Government.

Whilst the idea of collaboration for airframe manufacture would be aimed at Europe (e.g., Concorde), Plowden couldn't ignore the American market altogether given its market potential. Therefore, whilst looking to Europe for collaborative partners to design and build aircraft with governments effectively sharing the research, development and production costs, (something Plowden thought America would not be interested in given the size of the British market at the time); he recommended that complete units–engines and

6 Disappointly for Rolls-Royce, the contract for the Concorde Engine was awarded to Bristol Siddeley Engines.

equipment, should be developed and offered for sale to fit American airplanes. The possible tensions this might create did not seem to occur when making this recommendation.[7] An example of this outlined below, was the proposed collaboration between Bristol Siddeley Engines (BSE) with engine manufacturer Pratt &Whitney (P&W).

Rolls-Royce believed that it could offer a three-shaft version of the RB178 at a much higher thrust than the demonstrator engine and advised the Ministry of Aviation of such in February 1966 as both Boeing and McDonnell Douglas had invited Rolls-Royce to quote for a 'big' engine with a high bypass ratio for a long-range, wide-bodied aircraft with passenger and freight applications.

Elimination of a competitor – Bristol Siddeley Engines

This period of change on a political level was mirrored by the structural changes then under consideration within both the United Kingdom aircraft industry and, at a micro level, the changes considered and implemented by Rolls-Royce as a company. As a major manufacturer of aero engines, in a market forecasted to show worldwide growth in civil aircraft, Rolls-Royce in 1966 started informal discussions with the Bristol Aeroplane Company Ltd. and Hawker Siddeley Group Ltd, the shareholders of BSE, to acquire and place BSE into the Rolls-Royce group of companies. BSE was considered a major domestic engine competitor. For some time, Rolls-Royce had been disappointed with the overall level of government contracts placed with them, including the decision to place the engine order for Concorde in 1961 with BSE for its 'Olympus' engine, in collaboration with French company, Snecma.

When in June 1966, it had become apparent to Rolls-Royce, that BSE had signed a Heads of Agreement with P&W, Rolls-Royce's American competitor on large engine projects, with an option by BSE to take up licenses to manufacture the JT9D engine in the UK and thereby give P&W a foothold in the European market, this potential collaboration was viewed as a step too far by Rolls-Royce who purchased BSE, at a cost of some £63.6 million pounds, £21.9 million pounds being paid for 'goodwill', just over a third of the total price. Reproduced in the MacCrindle and Godfrey Report, Ibid. (1973) (*pgs. 21-22*) were 19 reasons and benefits of the proposed acquisition, which had been outlined in a contemporary document-part of the offer to purchase BSE, in particular, having one single aero engine manufacturer in the UK was

7 "The lure of the large and profitable American market could not be entirely denied and Plowden clearly saw no danger that this might create tension between a European Strategy for airframes and a transatlantic approach for other sectors of the industry". Ibid, (Hayward K. 1983), (*pg.73*)

compelling. Rationalisation and merger of the two companies, was considered in line with Plowden and the government's policies for the aircraft industry and would lead to economies of scale, less duplication of effort in design and manufacture and assist in the forming of future aero engine collaboration projects with European countries.

Although the Plowden Report recommended the sale of engines and products to fit American aircraft, the prospect of production of American engines being produced in the UK prompted this defensive move, preventing a potential collaboration of BSE with the American firm P&W and denying them a foothold in both Britain and Europe. Sir Denning voiced concern that through BSE's association with Snecma and the Heads of Agreement with P&W, it could give rise to the possibility of P&W's JT9D engine becoming the engine choice for a European designed airbus, then under discussion. It would have meant that the JT9D would have been in direct competition with the Rolls-Royce engine which was being aimed at the American market and whose preliminary development was being financed by the UK government. Conversely, Lockheed considered the potential for sales of the L-1011 in Europe and countries associated with the UK (See Chapter 9 pg. 285).

EVERYTHING STARTS WITH A RESEARCH PROJECT
The Work of the Rolls-Royce Advanced Research Department

> *"Ye see Smith, a jet engine is really quite simple, no matter what these chaps tell ye, there's a cold end where the air comes in and there's a big fan going round an there's an ot end, where the air is pushed out there's another fan going round. In between the two there's a fire in the combustion chamber to make engines better. We've got to make the cold end lighter, because it's too heavy, and the ot end otter, cos that makes it produce more power, and we've got to make the fire in the middle more efficient. That sums it up, that's all there is to it."*
>
> **Lord Hives, Rolls-Royce, Chairman in conversation (local dialect) with Sir Alex Smith, Head of Advanced Research, and Adrian Lombard, Engineering Director (1960)**

No longer owned by Rolls-Royce, The Old Hall, Littleover, Derby (1957 to circa 1971) was where some of the very early aero engine technology was developed, along with other emerging technologies such as composite materials, the electro jet process for drilling turbine blades and early engine propulsion for nuclear submarines, which Rolls-Royce would later go on to develop into a significant new business stream for the company.

Recollections of the Advanced Research department by Dr. Ian Macpherson, former Rolls-Royce Assistant Chief Research Engineer

"The Old Hall at Littleover, Derby, had been purchased by the company around 1957 when it became clear that the company's intention was to take part in nuclear power for a future developing business stream. This was certainly apparent to the Chairman, Lord Hives. A large laboratory was constructed with an associated machine shop, and a start made on getting significantly involved as a serious participant in the emerging nuclear industry. Early efforts assessed possible fuel element configurations in a rig where samples could be heated electrically and cooled by domestic water. This involved letting the Derby power station know when the power was to be turned on, usually in the early hours of the morning.

Around 1960, Rolls-Royce and Associates (RR&A) had been established at the Raynesway, site for nuclear activities and The Old Hall site was re-designated as the 'Advanced Research Laboratory' following the rundown of the earlier efforts into nuclear activities. This major new development at The Old Hall became a reality following extensive discussions between Alex Smith, Lord Hives and Adrian Lombard. Alex, who had previously headed up the company's involvement in nuclear power, was asked to take a long-term view of other potential areas of research. This new activity was strongly supported by Hives who had personally driven the nuclear developments, convinced in his mind that the company must gain a foothold in this emerging technology. Lombard was already exploring advanced design concepts for using components made of composite fibreglass resin structures in aero engines.

The activities at The Old Hall got underway with Air Chief Marshal Sir Ralph Cochrane, of Dambuster fame, as titular head. Alan Griffiths had his office there, no longer at the Duffield Bank House, and was actively pursuing his vertical lift ideas. He was refining the flying bedstead and pursuing a concept for a supersonic passenger aircraft (HSA-1011) well before the Concorde project, using multiple small vertical-lift engines for take-off (a concept that became a reality in the RB108 engine (RB162 in due course) and conventional jet engines for propulsion. Stanley Cooper, a solitary expert on bearings with one laboratory technician, was developing air bearings. At the same time, he was being called on by the aero engine division to help resolve their own bearing problems, for example with a groundbreaking design of the RB211 engine which had three concentric shafts. Another senior engineer was proposing rail cars using the diesel engines made at Shrewsbury.

As head of the laboratory, Alex Smith set up a small group to explore relevant long-term areas. I was asked to take on advanced gas dynamics and other possibilities as hypersonic combustion and magnetohydrodynamics, two emerging areas of interest in academia. The team included Jock Hill, a wizard when it came to electronics, instrumentation and the high-speed photography needed to capture Mach 5 shockwaves penetrating high strength magnetic fields. Brian Edwards eventually had a rig in which a hypersonic detonation wave could be sustained indefinitely in a circular 'racetrack' as an incredibly intense (and ear-splitting) form of combustion.

John Morley headed a larger group of resources on advanced materials. One of their early successes was to produce laboratory scale samples of incredibly strong glass rods. These were produced when reference was made to some very early work by Alan Griffith when employed at Farnborough. By firing 303 bullets at several empty gas cylinders, all of which that had been subjected to many cycles of repetitive pressurising and release in commercial work; Alan deduced that the unpredictable shattering of the cylinders occurred because of minute surface fatigue cracks. John Morley formed composite glass rods with two glasses, one with a higher melting point in the core and the lower melting point version formed around the core. Samples were about 15 cm long. Heating this assembly effectively annealed the central material free from surface cracks, with a continuous transition to the softer outermost layer, in a truly amorphous form free from discontinuities. When cooled the rods showed extreme strength and stiffness, capable of sustaining high levels of elastic bending. These properties were totally lost if a fingertip touched the outer surface. Eventually by using a cross bow in which the bolt was attached to one end of these heated composite rods and fired along the length of the laboratory, long lengths of the strong fibres were produced. About this time news came through that Farnborough was experimenting with carbon fibres. The efforts at Old Hall then turned to producing significant amounts of carbon fibre. This involved taking skeins of Terylene fibre before these were spun into thread, and "cooking" these in an inert atmosphere with the aim of driving off all the hydrocarbons to leave pure carbon fibres. The "furnaces" were metal tubes about 20cm diameter and 15m long and the inert atmosphere, nitrogen. Alex Smith describes the frustration and disappointments in achieving significant amounts of carbon fibre in his autobiography. Given a supply of fibre, further work on the laboratory bench scale showed it was possible to use the fibre to reinforce aluminium, the samples looking rather like a fibre shaving brush with a reinforced aluminium handle. When Adrian Lombard saw these structures, he immediately saw a possible application of using resin and carbon fibres, now called Hyfil, to make fan blades in aero engines. In short, a massive effort was put into effect to generate carbon fibre and form Fan Blades at a mini factory set up on Alfreton Road and at Hucknall to lay up these blades, using formers following the fibre glass techniques used in boat building. Using Hyfil composites, the fan blades had distinct advantages over titanium in about eleven of twelve characteristics. For example: cost, fan profile for best aerodynamics, lower fan weight, lower weight in the fan hub, lower bearing specification, ease of manufacture. Others are listed in the records. These were so attractive that the programme intent was to use Hyfil blades in the first batch of production level RB211 engines for the Lockheed contract, with titanium as back-up."

In his own words, the following are extracts from a paper dated November 23rd 1971 by Sir Denning Pearson regarding the Rolls-Royce company strategy, which was developed and adopted for the RB211 aero engine project in the 1960s which was the main focus on which the future prosperity and long-term financial security of the company was seen to rest.

"THE RB211 COMPANY STRATEGY [CIRCA 1960]

The company's strategy has been unchanged since the end of the last war. It has consisted of endeavouring to ensure that profitable and increasing use was made of the company's total assets of capital, people, accumulated knowledge and goodwill as the means of maximising the long-term return to the shareholders and fulfilling its obligations to employees, customers, suppliers and the community.

To implement this strategy the following tactics were adopted:

1. Keeping abreast of all scientific knowledge and new developments which might make existing products obsolete and carrying out the necessary research and development needed to keep them up-to-date and competitive, both in design and manufacture.

2. Keeping close to customers and the market requirements to ensure that new products were developed when the market required them and not before.

3. Keeping a constant lookout for opportunities to exploit the company's technical capabilities, production capacity and market coverage.

4. Examining all new projects for profitability in the context of project profit plans which covered all aspects of the projects over their full life. In the case of civil aero engines this was usually taken as eighteen years.

It was by pursuing this strategy that the company's commercial business had been built up since the last war to replace the continuous falling-off in military aero engine requirements."

"THE DECISION TO PROCEED WITH THE RB211 ENGINE PROJECT

Despite early success with the Dart (engine) which resulted at one time in more than 50% of the world's gas turbine-powered civil aircraft being powered by R-R engines, the virtual replacement of the propeller turbine by jet engines gave our American competitors the opportunity to adapt their military jet engines for civil purposes. Although we retained a substantial share of the market, they quickly overtook our lead. We were also aware that 85% of the world's airline aircraft were being produced in America and 65 to 70% were being put into operation by US airlines. It therefore appeared obvious to us that if we were to stay in the civil engine business which we had built up since the war, we needed to get into a US-built airframe as first engine choice.

In addition, about this time market surveys, confirmed by the Min.Tech's[8] *own assessments, showed that by 1985 some 80% of the world's subsonic civil aero engine requirements would be for engines of 40,000 lb. thrust or over. By 1966 we had settled on the three-shaft configuration and commenced preliminary discussions with the US airlines who were beginning to think of a new medium-range aircraft for domestic operations. We were at the same time pursuing the selection of an R-R engine for the European Airbus. In the event, the order for the RB211 for the Lockheed TriStar was finalised long before a decision was taken on the European Airbus.*

8 MinTech – the Ministry of Technology formed after the dissolution of the Ministry of Aviation by the Labour Government.

The risks involved in such a project were formidable. The step forward in technology was substantial, quite apart from the problems arising from the increase in size of the engine over and above anything we had done previously.

We let the government know that we were not prepared to find more than 30% of the launching costs as we would not risk a greater amount of the shareholders' money on a single project on which we are also responsible for the overrun. We were only able to go this far because of the merger with Bristol Siddeley which increased our base.

Why did we need to do the RB211 at all? Basically, because we saw no other way of continuing to make profitable use of our assets. Military orders were declining and likely to continue to do so. A question mark hung over the engine for the Concorde, and we could see no other alternative civil project. We also judged that new commercial engine projects would happen less frequently and that the opportunity for the RB211 might well be the last for some time.

The size of the gap needing to be filled is best illustrated by the numbers which I would estimate today are employed on the RB211 project – one half of the total engineering staff of the Derby Engine Division of approximately 2500; two-thirds of the Derby Engine Division experimental manufacturing staff of about 6000; one-third of the total of about 18,000, employed in the production organisation of Derby Engine Division, and a total of well over 20,000 people, including those working on bought-out and sub-contracted parts.

Quite apart from the RB211 itself, I am quite certain that the UK would not have succeeded in getting the contract for the engine for the European multi-role combat aircraft (MRCA) (Tornado), without the existence of the RB211 project. It was only the existence of advanced technology at R-R which was exemplified by a US constructor and the US airlines choosing the RB211 that persuaded the German government to accept a British engine for the MRCA project in the face of intense American competition. Without this project there would have been a further gap at Bristol and under utilisation of physical and human assets. It could, and has been, argued that the people and assets now being used on the RB211 and the RB199 for the MRCA project could be put to better use on other work.

The simple answer is that we could not foresee any alternative projects. There just was no pressing world demand for other new civil engines for new aircraft projects. One possible alternative could have been civil STOL or VTOL[9]*, but these are still not being financed by the British government some three years later. R-R have suitable engines in the project stage; but there are no signs, even today, of finance for their development. In view of this, it is inconceivable that the government would have been financing today any alternative project that would have employed a fraction of the physical and human assets now being used on the RB211 and RB199 [programmes].*

The company's fixed assets of plant and machinery are highly specialised and largely useless for anything but aero engine work, so that it would have been difficult to see the company continuing as a viable company on a long-term basis if the utilisation of its assets had been reduced by reduction of the number employed by those now working on the RB211 and RB199.

These issues were under continuous discussion by the Main Board from 1966 onwards and when the time came for a final decision the Board agreed to go ahead, subject to Government participation and on the basis of the Project Profit Plan which showed a satisfactory return on capital employed." . . .

Sir Denning Pearson
(Former Rolls-Royce Chairman, November 23rd, 1971)

THE US CIVIL AIRCRAFT MARKET AND THE RB211 ENGINE OPPORTUNITY

Recollections of Richard Turner, former Rolls-Royce Commercial Director and Main Board member

The RB211 Engine and its Competitors

"It is important to recognise the origins of the RB211 engine which morphed out of Geoff Wilde's Preliminary Design department.

The first large three-shaft design was the RB178 which was the Rolls-Royce (unsuccessful) proposal for the original Boeing 747 aircraft sales campaign. The B747 inaugural Pan Am flight to London Heathrow resulted in two engines being removed there and Joe Sutter, the creator of the B747 'minor' was not amused! I remember the announcement that Rolls-Royce had lost the B747 competition when I was a graduate apprentice (in the mid 1960's) working on

9 STOL – Short Take-off and Landing; VTOL – Vertical Take-off and Landing

attachment in Administration, now called Human Resources, on Sinfin B Site, Engine Assembly Shop. At the time, there was actually another three-shaft engine under development called the Trent, about 15,000 lb. thrust for the projected Fairchild FH228 aircraft but this died when the aircraft was cancelled.

After the inglorious failure to convince Boeing, attention moved to the specification devised by Frank Kolk of American Airlines for a New York–Chicago route, which required wide-bodied twin-engine planes. Rolls-Royce offered the RB207 engine at about 50,000 lb. thrust. This was also offered to Airbus which was at that time creating the A300 aircraft. It must be remembered that this was before the deregulation of the US airline industry when competition was very much discouraged and the skies were dominated by the 'big four' domestic carriers, American Airlines, United Airlines, Eastern Airlines and Trans World Airlines (TWA) with Pan American restricted to international routes with some competition from TWA. Delta Airlines was a growing force but smaller than the big four.

As Boeing had its hands full developing the B747 with, as it seemed, very little help from Pratt & Whitney, McDonnell Douglas Corporation (MDC) and Lockheed entered the competition.

MDC had a successful track record with the DC-8 and DC-9 aircraft whereas Lockheed had withdrawn from the civil market after major problems with the Allison Engine Company powered, Electra turboprop. Whilst it did rectify this problem, the aircraft never recovered its market, which in any case was being taken by the early DC-9s and B737s. Ironically, Lockheed went on to make hundreds of Electra military derivatives as the US Navy P-3 ASW (anti-submarine warfare) aircraft which was still in production in Burbank when the L-1011 was being manufactured in Palmdale. The P-3 is still in service, and Lockheed profited mightily from this, as did Allison, together with powering the C-130 Hercules aircraft out of the Lockheed Georgia Plant.

However, Lockheed, led by Dan Haughton as Chairman and Carl Kotchian as President–both of whom I knew well, were determined to get back into the civil market and thus the DC10/L-1011 competition was born.

As the airlines started to concentrate on the two aircraft, they quickly morphed into machines with coast-to-coast rather than New York–Ohio valley range. Engine reliability also dictated (thank heavens) that twins were ruled out and tri-jets were created. The travails of the JT9D engine in Seattle meant that Pratt & Whitney (P&W) was swiftly ruled out of the competition, which was again ironic as it then had about 95% of the civil jet engine market with the Boeing 707, 727, 737 and DC-8 and DC-9 airplanes.

The propulsion competition therefore came down to Rolls-Royce and General Electric (GE). The GE proposal was based on the core of its TF39 engine, developed at US government expense and already in service on the Lockheed C-5A (5B & 5C), Galaxy military transport. GE's only previous civil engine had been the rear-fan engine on the Convair 880/990 which had sold very poorly and took Convair out of the commercial market.

The Rolls-Royce proposal was the RB211 which was a scaled-down version of the RB207.Then the battle took place–Lockheed versus MDC and GE versus Rolls-Royce!

Rolls-Royce and Allison (then part of General Motors) had enjoyed a successful relationship with the licence built TF41 'Spey' in the A7 attack fighter. The original RB211 proposals were a joint Rolls-Royce/Allison alliance but the airlines soon indicated that they didn't want any part of Allison and this partnership was abandoned. A side effect of this decision was that Rolls-Royce no longer had an engine to offer Airbus (then a very small enterprise) for the A300 which ruined relations between the two companies for many years. Airbus had to wait to launch the A300 until GE produced the CF6-50C growth version of the original CF6 for the DC10-30."

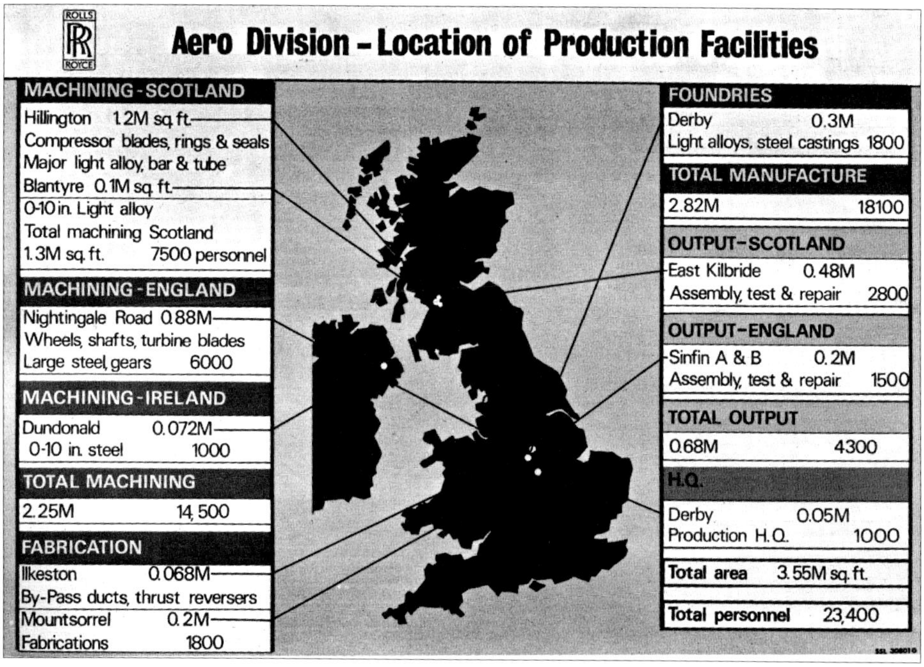

19. Rolls-Royce production facilities in the UK, 1968.

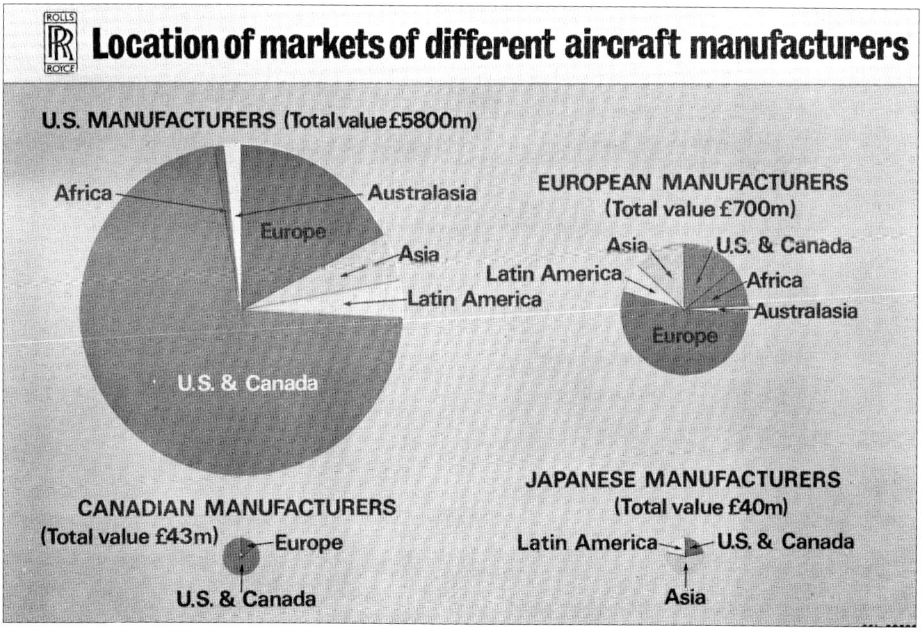

20. Rolls-Royce chart–location of world aircraft manufacturers, 1968.

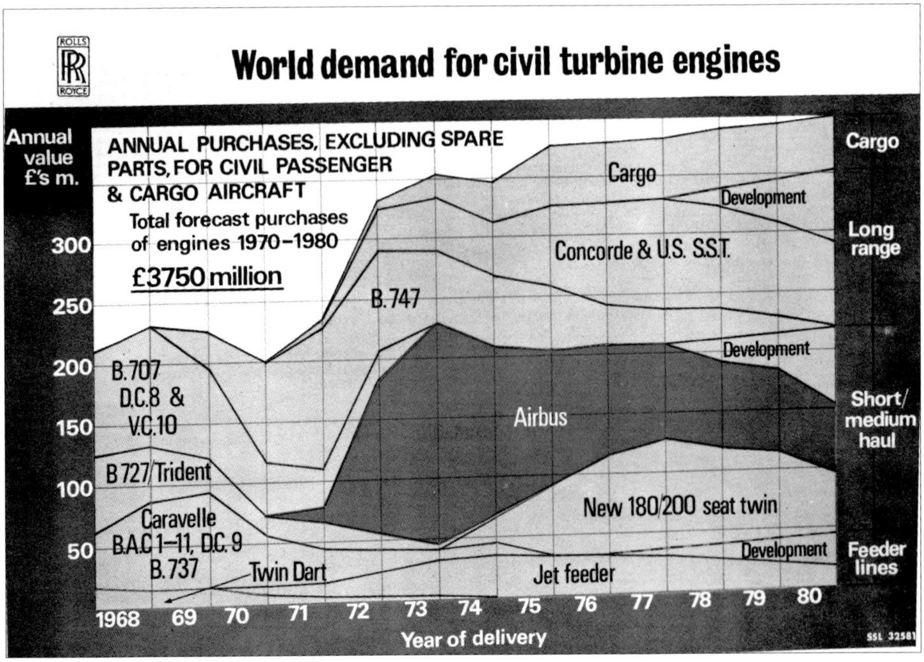

21. Rolls-Royce chart–world demand for civil engines, 1968.

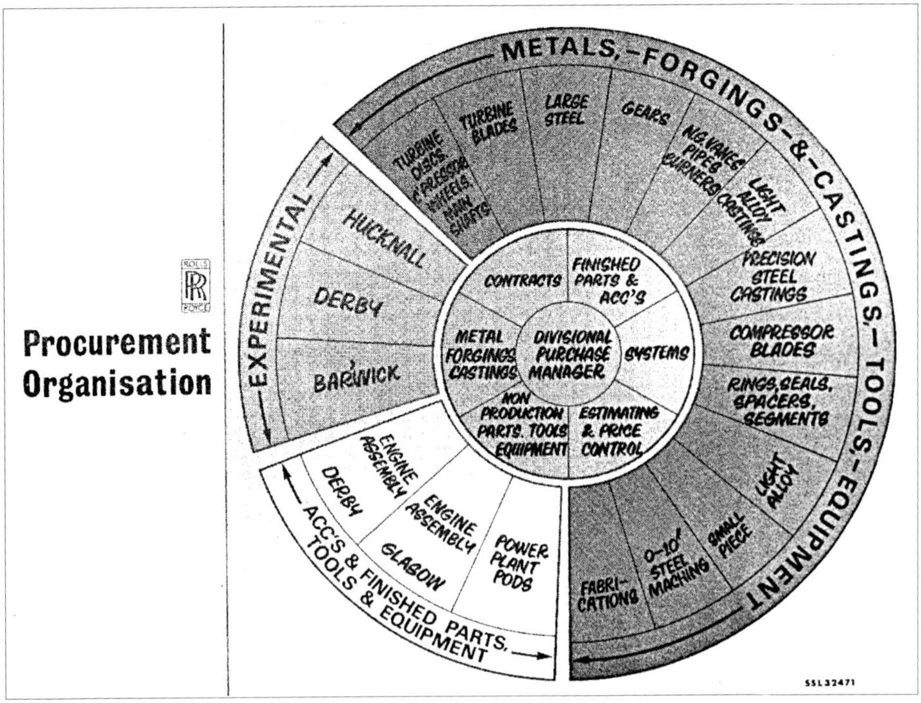

22. Rolls-Royce purchasing organisation in 1968.

COPYRIGHT ACKNOWLEDGEMENTS

Images 19–22 all copyright Rolls-Royce Ltd. taken from the Suppliers Conference Brochure, Solihull 1968. (See next chapter).

CHAPTER 3

THE £1,000 MILLION ORDER

"Rolls-Royce still had to convince the airlines and so stepped up its already frenetic sales campaign to an even higher pitch. David Huddie, later knighted for his part in capturing the contract, left his post as managing director of Rolls-Royce aero engine division in Derby to set up house and office in New York. The campaign which he directed was a hard-hitting, all-out affair, owing more to Madison Avenue techniques than to the traditional Rolls-Royce image of sleek, discreet, gentleman's motor cars with the famous silver lady on the bonnet. Rolls-Royce's executives raced backwards and forwards across the Atlantic liaising with aircraft manufacturers, airlines and government departments.

Sales literature was churned out by the hundredweight; telex and telephone messages chatted at any hour of the day and night. 'Brochuremanship' was never brisker. Rolls-Royce's publicity of the time showed . . . explanatory books piled higher than a man. They include a social and economic history of Britain since the war, and a black-bound folder, the gold lettering on the spine of which stated challengingly, 'Reasons why Eastern Airlines should choose the Rolls-Royce RB211 engine for their US tri-jet'.

At the end of it all, the Rolls-Royce campaign in America had cost £350,000, of which airfares and expenses accounted for £80,000. British-based staff had made 230 journeys to the United States and US based (personnel) a further 500 visits to airlines and airframe companies. Transatlantic telephone calls from the Rolls-Royce factory at Derby totalled over 100 hours and cost more than £6000, while telex messages were measured in hundreds of yards. The onslaught produced results. Two American airline giants, Eastern and Transworld, signed up for the Rolls-Royce-powered Lockheed airliner and on Friday, the 29th of March 1968, at a press conference replete with euphoria at the Connaught Rooms, London, Sir Denning Pearson, Chairman of Rolls-Royce, was able to announce the deal."

Arthur Read
The British Aircraft Industry: What went right? What went wrong?
J. M. Dent (1973)

At the time of the above announcement held simultaneously by Denning Pearson in London and David Huddie in New York on the 29th of March 1968, the RB211 was forecast to be Britain's biggest ever export order, certainly the biggest commercial sales order in the company's history and indeed the biggest in the British aerospace industry at the time. With Lockheed's prediction of future potential sales, it was quoted in the Observer Newspaper as being expected to boost Britain's balance of payments, (over the life of the contract), to the tune of £1,000 million. (*Pugh P. (2001) 'The Magic of a Name, The Rolls-Royce Story Part two: The Power Behind The Jets', Icon books UK, pg. 120*). The press release was published in the Rolls-Royce News on the 10th of April 1968. Both Denning Pearson and David Huddie were subsequently knighted for their roles.

The Institute of Marketing, Leicester Branch gave a dinner on the 17th of June with the following guests of honour, responsible it stated, for securing the order and to celebrate the contract ; John Bush, Engineering Director-Large Transport Engines, Rex Nicholson, Production Director, Ernest Eltis, Director of Engineering, Tom Metcalfe, Production Control Director, Mr. Wheatley, Manager Production HQ and Alan Newton Assistant Contracts Manager, June 1968 (Image 24).

My RB211 Sales Campaign Experiences from the man who was there - Further commentary by Richard Turner :

"My recollections cover a period of over 50 years and with the passage of time may be a little blurred and you must also appreciate that I am not an engineer but a graduate from Manchester with a very poor 2:2 in Politics and Modern History, a qualification that wouldn't even get me through the door if I was applying to Rolls-Royce today. That qualification later came in handy when travelling the world for Rolls-Royce, especially when I visited Russia many times in the mid-1970s in the later days of Brezhnev. Communism in action didn't bear much relationship to Marxism!

In those days, Rolls-Royce had a policy of recruiting 15% of graduate apprentices as non-technical people and, against the odds, a number of us did make a contribution. The graduate intake in 1965, the year I joined, was about 100 in Derby alone and this was before the merger with Bristol Siddeley and in the days when only about 10% of school leavers went to university.

I was personal assistant (PA) to David Huddie, from about 1965 to about 1968. Huddie always had a PA from the non-technical graduate apprentices, and I was lucky enough to be selected. The job was really a general dogsbody, but it did enable you to see the whole company from the very top and this perspective was very useful when I left the position. The general rule was that the job was only for one year but the pressure on us during the efforts to launch the RB211 was such that I lasted three years.

David Huddie's office was led by Winnie Buchan as his secretary. Winnie was a strong character who stood no nonsense from anyone. She had previously worked for Adrian Lombard until he died. There were three staff officers covering the engine projects. Ralph Robins did the Spey and TF41 military version, Tony Manners covered the Tyne, Dart and older engines and Derek Taulbut covered the RB211 and wrote all the launch aid applications to HMG – of which there were many! Peter Colston, an executive assistant, moved on and was succeeded by Ralph Robins, with whom I have remained friends ever since.

Huddie moved to New York to run the campaign and we supported him from Derby. This usually involved being in the office until late evening as we reacted to events taking place in US time.

When the campaign started, the RB211 was offered to both Lockheed and MDC and 'launch conditions' were agreed with both manufacturers that the engine would only be launched by orders from three of the big four US airlines-American, United, TWA and Eastern. This was because it was made clear that Rolls-Royce could only manage one aircraft programme.

Both Lockheed and MDC accepted this, and legal agreements were signed with each of them. I have no proof but believe that GE had the same agreements.

In addition, these provisions meant that, in the end, only one aircraft and engine would be launched and implicitly this recognised that the market was only big enough for one aircraft – which history proved correct now that MDC is part of Boeing and no longer making commercial aircraft. Lockheed is similarly out of civil aviation and Airbus has captured its market. I don't think either the L-1011 or the DC-10 made any money for their companies. As the campaign proceeded, Huddie and his team concentrated on the airline launch customers and managed to convince them – unlike the internal Lockheed selection team whose engine choice was GE's CF6. The RB211 still had the Hyfil blade engine at the time. Now, I can't remember the exact dates, but the climax started when American Airlines, which had issued the original specification, decided unilaterally to select the DC-10 with the RB211 and planned to announce this on a Monday. However, on the preceding Saturday, C. R. Smith, who was both the founding father of American Airlines and its Chairman, was appointed US Secretary of Commerce by President Lyndon Johnson. In these circumstances, American felt unable to announce that it was buying a non-American engine, so the announcement stated that the engine selection would be made later. Nevertheless, American agreed a side letter with MDC saying that the RB211 was the 'preferred power-plant'. I think, but cannot prove, that Rolls-Royce got a copy of this.

When, Haughton (Lockheed Chairman) heard this, he got hold of Huddie and said that, if Lockheed was to agree to launch the L-1011 with only two of the big four, would Rolls-Royce agree to change its launch conditions. Obviously, I was not party to the internal discussions within Rolls-Royce and the Board, but the upshot was that Rolls-Royce agreed and Lockheed swiftly announced the launch of the L-1011 with Eastern and TWA, ironically the two financially weakest of the big four. Delta, considerably smaller then than today but growing strongly, followed a few days later and Air Canada joined sometime later. An order for, I think, 50 aircraft from a hastily created British company called Air Holdings was also announced in the hope that BEA (British European Airways) would order the aircraft. In fact, BEA did so some years later but, by this time, the Air Holdings facade had disappeared!

This left both MDC and American with an unlaunched airplane and considerable bitterness, feeling that Rolls-Royce had acted in bad faith by altering its launch conditions on which both companies had relied. Phil Gilbert, head of the New York law firm that Rolls-Royce used, immediately told the company that it was liable to anti-trust proceedings by only offering the engine to Lockheed. Accordingly, an emergency meeting was arranged in Derby with MDC to demonstrate that Rolls-Royce had the resources and capability to do both aircraft programmes.

An MDC team led by Dave Lewis, the CEO and a very nice man who got on very well with Huddie, came to Derby. I sat in on some of these meetings which were dismal affairs as everyone knew that in reality it was a load of rubbish. Inevitably, MDC didn't believe this but there was no movement for some months until United agreed to buy the DC-10 with the only realistic engine on offer, the CF-6. This launched the DC-10 programme.

In retrospect, if C.R. Smith had not been appointed Secretary of Commerce that weekend, American would have announced the DC-10/RB211 simultaneously and, in these circumstances, I suspect Rolls-Royce wouldn't have agreed with Haughton to alter the launch conditions and there would actually have been only one programme which would probably, even given the Rolls-Royce receivership, have been a success! The rest is history. Ironically, American subsequently became Rolls-Royce's largest customer in the world with large fleets of B777s, B757s, Fokker 100s and hundreds of Embraer Regional Jets -ERJ145.

A significant issue at the time of the receivership was that, if the L-1011 had been cancelled, the deposits that the launch carriers had made would have been non-returnable and this would almost certainly have put the two most financially weak airlines (Eastern Airlines and TWA) into bankruptcy. I think this was a major factor that persuaded the US government to meet the UK government's stipulation that Lockheed must be given a US government guarantee. As is well known, this passed in the US Senate by only one vote and was steered through by Senator John Connally (ex-governor of Texas). Ironically, he was the man sitting next to President John F. Kennedy in the presidential motorcade in Dallas, Texas, when Kennedy was assassinated in 1962.

After February 1971, I was quite involved in the relaunch of the programme as I had by then been moved to the North Hollywood office and only got eight nights in my own flat between April and September 1971.

As a postscript, in his book, "The History of the Rolls-Royce RB211 Turbofan engine", *Phil Ruffles covers the very difficult negotiations with BA (British Airways) to launch the RB211-524 in the B747-400. "I was very much involved in this life-or-death competition as failure would have put Rolls-Royce out of the big engine business for good and given GE a near monopoly. It did entail breaking the spirit, if not the law, of the agreement with GE and, again, led to much bitterness. I was one of a number of people who had very unpleasant meetings in Cincinnati".*

Phil Ruffles also refers to the poor relationship between our then Chairman, Lord Tombs, and Lord King, Chairman of British Airways. As I remember it, King stood up in the House of Lords and announced that Rolls-Royce was "only fielding its second team". I was among a number of us who were incensed by this remark, and I won't go into detail about how we broke the deadlock with BA in the gent's loo in the Rolls-Royce office at Heathrow during the very rough negotiations held over a weekend!

However, when we finally won, at great cost, but ensuring that we survived as a civil engine manufacturer, I commissioned 30 celebratory T-shirts which we gave to all those involved in the struggle with BA, and you may be amused by the below."

T-shirt

COMMENTS ON THE ORIGINAL RB211 CONTRACT BETWEEN ROLLS-ROYCE AND LOCKHEED

The following are extracts from comments on key contractual terms contained in the original RB211 contract between Rolls-Royce and Lockheed made by Sir David Huddie from a paper dated 20th November 1971.

"Since February 1971, much has been written and said about this contract and the following summary of some points in the contract might be helpful. The comments are under four headings as follows:-

1. PRICES

a. Prices were fixed for deliveries up to the end of 1971. The number of power units which could have been delivered before the end of 1971 was limited to a maximum of about 34 aircraft sets by capacity and maximum acceptable rate of build-up of production deliveries.

b. Prices for 1972 and 1973 deliveries were fixed subject to escalation to a formula covering wage rates and material costs.

c. Prices for 1974 and 1975 were subject to new quotations and escalation, except for power units ordered or on firm option by launching customers by a date.

d. If at any time Lockheed raised the basic price of the airframe, Rolls-Royce had the right to raise the power unit price by the same percentage.

The commercial aircraft and engine business has traditionally been highly competitive, and there cannot have been many occasions when manufacturers have been satisfied with prices for launching orders. In the past several methods have been used to correct or partially correct this initial situation.

A. Prices have been raised in later quotations. This feature has applied to engines and aircraft.

B. Most commercial engines and aircraft have had developed versions. Such developed versions gave clear opportunities, which were taken, to quote higher and more profitable business.

C. Spare parts are more profitable than complete aircraft or engines. This is more important in the engine business than in the airframe business.

While some of these features were very unwelcome from the manufacturers' points of view, they were part of the framework within which the RB211 competition was carried out.

At the time of signing the contract, Rolls-Royce judged that the profitability of the total project was satisfactory. It was also the judgement of my colleagues who were closely concerned in the negotiations that the terms obtained were substantially as good as could have been obtained at that time.

The conditions of the Rolls-Royce/Lockheed contract were, to a significant extent, passed on by Rolls-Royce in their contracts with major suppliers. These contracts were cancelled by the bankruptcy of Rolls-Royce.

2. *ESCALATION*

Escalation was by a formula covering wage rates for the aircraft industry and the prices of materials and fuel used in Mechanical Engineering Industries using indices then published regularly by the Department of Employment and Productivity.

3. *LIABILITY FOR LATE DELIVERY*

This clause was put in to protect Rolls-Royce by limiting its liability for late delivery. While no limited liability clause can completely protect a supplier from its liabilities under common law, it was the opinion of the Rolls-Royce lawyers in the USA that this clause in this contract would almost certainly be upheld in the courts.

Causes for late delivery were in two categories- those outside the control of Rolls-Royce and those within the control of Rolls-Royce.

A. Causes Outside the Control of Rolls-Royce

The list of such causes was wide and included action by any Government causing delay, work slow-downs as well as strikes and civil disturbances, as well as war, fuel, power and material shortages and modifications required to meet regulations by the Air Registration Board in the United Kingdom or the Federal Aviation Agency in the USA, after the date of order. Rolls-Royce had no penalty for delay under this heading, but Lockheed had the right to cancel items delayed for more than six months provided they gave notice within 30 days after the six months.

B. Causes within the control of Rolls-Royce

For such causes, Rolls-Royce was required to pay for air shipment of delayed supplies and, should Lockheed suffer damages nevertheless, Lockheed had the right to recover liquidated damages for delayed supplies at a rate using from ½% to 2% of the purchase price per month for a period of seven months, up to maximum of 11%. For delays beyond seven months, Lockheed had the right to cancel by giving notice during the eighth month of delay.

4. GUARANTEES

The contract contained several guarantees on performance and weight. These guarantees were tight and in the event all of them would not have been met. The contract provided that, in the event of failure to meet one or more guarantee, the guarantees should be judged in aggregate and that any failure to achieve specific guarantees would be considered remedied in whole or in part by achievement in excess of guarantees established in other areas where such improvements over guarantee, offset the results of deficiencies, always provided aircraft guarantees and performance were not adversely affected.

Argument about failure to meet specific guarantees is always difficult, but at least acceptance of the principle of offsetting guarantees against each other helps.

In particular, the situation where the Hyfil fan blade might not work or be ready in time was recognised and weight and performance penalties of fitting an alternative titanium fan blade, should this be necessary, were covered."

SELECTED EXTRACTS FROM A SPEECH GIVEN AT THE ROLLS-ROYCE RB211 SUPPLIERS CONFERENCE IN SOLIHULL AND FROM THE SUPPLIERS CONFERENCE BOOKLET JULY 1968 AND THE ROLLS-ROYCE 'REPORTER' MAGAZINE 1970

This conference brought potential sub-contractors and suppliers together and an outline of what the contracts would entail and what they would be expected to comply with, some of which are set out below:

There would be no recovery from Rolls-Royce of any cost for a change in goods or services unless the supplier could show the cost exceeded 10%. They would be given contracts drawn up to include many of the terms of the Lockheed /Rolls-Royce contract for example, for orders received on or before 31 December 1971 prices would be fixed, after this price increases would be based on an escalation formula as per the same material and wage indices as in the Lockheed contract. However, there would be no extended credit terms. Delivery of goods to Rolls-Royce would be strictly within the agreed programme and if delayed the supplier would be liable to pay liquidated damages of 2 % of the price of the goods for each week of delay up to a maximum of 12 %. There would be warranties and liabilities for failure of parts and suppliers would be responsible for the cost of replacement if they resulted from faulty 'design, material or workmanship'.

REX NICHOLSON (Rolls-Royce Former Production Director)
The RB211 Contract: Rolls-Royce's Expectations of Suppliers

> *"Since July 1967, all areas of the aero division (engineering, commercial, experimental, manufacturing, sales, service and production) have been involved in selling the RB211 engine. The Rolls-Royce sales effort succeeded, by demonstrating the total involvement of all company departments within the aero division, in setting a programme selling price.*
>
> *The price was arrived at from a production cost estimate which committed us to a material and accessory cost, labour content and the total cost per hour in 1971.*
>
> *We have no opportunity to correct any assumptions which we made in any of these areas and recover the difference from the customer. We must, therefore close the design changes to meet the specification, and find ways of maintaining the cost quoted. Should total costs per hour exceed the estimates which we have used, we must take steps to reduce these by whatever means we can or suffer the consequences.*
>
> *You will hear that we must, therefore, expect the same attitude of mind from our suppliers. The prize is for Rolls-Royce to continue in the business of supplying aero engines into the biggest aircraft market in the world. Failure would mean a rapid decline and contraction of the aero business.*
>
> *The rapid growth in air traffic has created problems in handling the large number of travellers on a worldwide basis. The reductions in operating costs to be obtained from the present generation of jet aircraft, i.e., Boeing 707, Douglas DC-8 and Boeing 727, have been almost fully exploited and the cost reduction curve is now bottoming out. The new generation of aircraft will soon be needed to restore the downward trend. Larger aircraft enable passenger seat-mile costs to be reduced.*

World demand for civil turbine engines and the value of new engine deliveries to the world's airlines now exceed £200 million per annum, including spares and associated equipment. It will increase to £350–400 million per annum in the next 10 years. Airbus aircraft will be the largest single market up to 1980, accounting for almost one-quarter of the total requirement.

The US market is important for us as US aircraft manufacturers supply over 80% of the total world need for aircraft and US airlines account for 60% of the total airlines' requirement."

THE RB211 CONTRACT: COMMENTS ON THE COMMERCIAL AND FINANCIAL FRAMEWORK – taken from the Supplier Conference Booklet

J. B. PERKINS (Rolls-Royce Former Commercial Director)

"The RB211 propulsion system price has been fixed for a period of years ahead at an extremely competitive level for a very high standard of performance and was based on cost estimates today before design is complete and for tomorrow's advances in manufacture. This is a hard commitment but one which has become usual practice over several years past. In consequence, Rolls-Royce has developed their philosophy, organisation and procedures to fulfil such commitments profitably.

Rolls-Royce has quoted a firm price for 1971 production deliveries. This has been pre-escalated and has no escape clause on excess escalation. For deliveries in the years 1972 to 1975, Rolls-Royce have committed themselves to the same basic price, together with an escalation formula, based on national indices for those parts of the price appropriately related to labour and materials.

There is no provision whatsoever in these prices for subsequent adjustments due to the inevitable unnecessary process of design change which is associated with developing and meeting this certification of an advanced engineering programme of this time. All these have had to be predetermined and included in our estimate. We have also had to make substantial provision for inclusion of costs due to change in the FAA (Federal Aviation Authority) requirements on the propulsion system.

It is, of course, necessarily prudent in any programme of this nature for the company to make contingency allowances for design no matter how many fetching problems, and this we have done. However, at least at two stages during the negotiations it was necessary to make significant price reductions. In addition, we agreed to an undertaking to absorb a major proportion of any future increases in US import duties over and above the present planned Kennedy Round figures.[10]

Apart from our commitments on additional equipment, the company made undertakings on the maximum parts cost of the engine for periods of 10 years' operation in service. This places significant restraints on the future spares pricing policy of the company and the suppliers on the reliability of life of our products.

One of the contract conditions we have had to accept from the aircraft constructor is significant penalties for failure to deliver acceptable supplies at the agreed date. Direct penalties run up to 11% over seven years of the total propulsion system price and with the costs of airfreight of course as well borne by Rolls-Royce.

We have also had to accept a withholding clause amounting to seven percent on engine price associated with failure to meet specific performance. Normal comprehensive Rolls-Royce engine and parts have been given life warranties with commitments well beyond the undertakings that were previously made on earlier engines. For example, on the main shafts and discs we are giving a 12,000-hour warranty. It was necessary to enter into a rigorous termination agreement that has significantly limited our termination recovery in the event of any termination during the next 12 months.

In conclusion, the commercial terms we have agreed are what we have found it necessary to accept to acquire the order for the RB211. It is indeed the price it was necessary to pay to enter and remain in the competitive aero engine business of today. We believe we have a highly successful and effective engine which will succeed in this airplane and in this market."

10 The Kennedy Round figures was the 6th round of a total of 8, of negotiations within the GATT tariffs. The 6th round which took place in Geneva between 1964-67 was so named in honour of President Kennedy who had been assassinated in November 1963, six months before the start of the negotiations. This round was hailed as the most important negotiation achieving an 'across the board' or 'linear' approach to reducing tariffs by an agreed percentage for those participating countries and for certain commodities, as opposed to haggling over each commodity. (A more detailed description can be found in Norwood B. 'The Kennedy Round: A try at a linear trade negotiation', *The Journal of law and Economics*, Vol.12, No.2, Oct.1969, pages 297-319. (University of Chicago Press Journals).

A FINANCE DIRECTOR'S VIEWPOINT OF THE RB211 PROGRAMME'S FINANCIAL STATUS IN 1970

DAVID FLEMING, Former Financial Director, Derby Division 1970
(Extract from *Rolls-Royce Reporter Magazine*, May 1970)

"Speaking regarding the company accounts which were released in May 1970 and referring to the current financial situation and emphasising the effects of rising RB211 engine development costs. These must be diminished by increasing productivity and I want to make clear all these factors which concern the company, and which to my mind are part of the total approach which is broader in its implications than is generally realised.

It will do no harm to restate the facts concerning the Rolls-Royce company. It has assets of £221 million. At the end of 1969, it had 87,000 employees and 90,000 shareholders, which represents a fair balance, typical of large public companies.

The Shareholders

Taking first the shareholders, many of us employees are shareholders, either directly as holders of stock, or indirectly through the pension funds, insurance companies, or unit trusts. These 90,000 shareholders are the people who invest money which provides the company with funds to finance its day-to-day running, and particularly the development of its products and expansion of its markets which are so essential to its future.

Just as we as private individuals expect to receive interest on the money we invest in savings certificates, Building Societies, government securities and deposit accounts, so our shareholders expect a reasonable return on the investment in the company. They realise that historically this return will not compete with those of other industries because we retain a large part of the profits for financing future developments in this advancing and highly competitive business. Unless people can feel confident that they will see a reasonable return, they will not invest their money in the company, and we shall be unable to finance our necessary activities.

It is here that we as employees have a double self-interest. Both as shareholders and as employees we can only gain by increasing the prosperity of the company. At the same time, in order to have the necessary financial support, we must show the public which has money to invest that we possess the will and the means to raise the overall efficiency and the productivity and cost-effectiveness of the company.

Our Competitors

We are engaged in a particularly competitive type of business. Our competitors are relentless in taking advantage of every technique to increase their efficiency, to produce better designs and to reduce costs, in order to hold on to or increase their share of the available market. There can be no standing still in this business. To succeed we must offer to our customers better value than they can obtain anywhere else. This involves quality and reliability in their broadest sense.

We must all of us realise that each one of us, and all the tens of thousands of employees of the firms which supply us with goods and services, has an important contribution to make. This starts from the quality of design through development to all the phases of manufacturing and procurement to after-sales service. We can all work more efficiently and an integral part of realising this objective lies in reducing costs at all levels of activity.

I happen to believe in the innate common sense of the British people. I am convinced of the loyalty of Rolls-Royce employees, and I am sure by combining this with a measure of self-discipline they can form a team, the aggregate effect of which will beat our competitors in providing better value, that is better quality and reliability of goods and services at a lower price. This way our future and our prosperity will be assured. I do not think that there is any other way.".

23. RB211 sales campaign literature–by volume!

24. Institute of Marketing, Leicester Branch dinner 17th of June 1968 held for 'Securing the order and celebrating the contract'. John Bush, (Engineering Director-Large Transport Engines, Rex Nicholson, (Production Director), Ernest Eltis, (Director of Engineering), Tom Metcalfe, (Production Control Director), Mr. Wheatley, (Manager Production HQ) and Alan Newton (Assistant Contracts Manager).

CHAPTER 4

THE KEY PLAYERS – PEARSON, HUDDIE AND HAUGHTON

Sir Denning Pearson and Sir David Huddie, (DPH) Chairman and Managing Director of Rolls-Royce Limited respectively, together with Daniel Jeremiah (Dan), Haughton, Chairman of Lockheed Corporation, were the three principal characters in this unfolding drama. Sir Denning and Sir David Huddie's joint involvement and influence were paramount throughout the timeline of the early development of the RB211 engine through to the winning of the RB211 contract. Sir Denning Pearson and Sir David Huddie had already enjoyed similar, long and distinguished careers as engineers with Rolls-Royce and now in their roles as Chairman and Managing Director, their vision was for world, rather than just British markets and believed their overriding mission was to secure an order as first choice engine supplier, from an American aircraft manufacturer and gain a foothold within the US market; in their view, ensuring the company's future security.

Dan Haughton was intrinsically involved with Rolls-Royce and the RB211 for 10 years mid 1960's to mid-1970's. For Dan Haughton and the Lockheed management the design and build of a new wide-bodied aircraft was a way back into the civil aircraft market and the engine chosen to power that aircraft would be the innovative three-shaft, RB211 engine.

Sir Denning Pearson and Sir David Huddie's joint determination to succeed with the RB211 in the US market along with key decisions they took would lead ultimately to both the fall and rise of the company – within three years of winning the contract for the RB211 in 1968, the engineering costs of development had spiralled, exceeding all the estimates, such that the company had a major liquidity crisis which would force it into bankruptcy in February 1971; the RB211 was eventually developed successfully and would go on to become a very profitable engine for Rolls-Royce in later years. An indication of this at the end of 1991 saw that the total RB211 business won had amounted to "£11,500 billion in 1991 monetary values".[11]

11 Figure extracted from *Huddie D. Lecture to the Rolls-Royce Heritage Trust* in October 1992.

A CAREER AND PERSONAL PROFILE OF SIR DENNING PEARSON

(Rolls-Royce Limited: Managing Director 1954–1965; Chief Executive and Deputy Chairman 1957–1968, Chairman and Chief Executive 1969–1970).

An alumni of Trinity college Dublin, Sir Denning Pearson joined Rolls-Royce from Metropolitan Vickers in 1932 with impeccable engineering qualifications and a first-class honours degree. His career progress to the top of the company was not through the normal pathways of research, design or development. In fact, he was one of an elite group of selected people responsible for setting up the Glasgow 'shadow factories' for the production of Merlin engines during World War II (Shadow factories were a second source of production in case the Derby Main Works was bombed and unable to manufacture).

Prior to World War II, Rolls-Royce had no engine or foothold in the civil aviation engine market. It was Sir Denning's recognition of the need for it to develop new business and engine opportunities that led him to persuade E.W. Hives, the then Managing Director, to develop a civil application for the successful Merlin engine. Development of a commercial version of the engine, designed for civil operation on a modified Douglas DC-6 aircraft followed. Forty-four of these modified aircraft were sold to Canadian airlines as well as the Royal Canadian Air Force as well as BOAC (now British Airways). It was at this point of his career that Sir Denning then transferred to the newly formed Rolls-Royce Canada Company to continue the sales and services functions for these aircraft. Some invaluable knowledge and experiences were gleaned from its engine in-service operations which were applied to the production of a variety of gas turbine engines installed in civil commercial aircraft in the1950s, such as the Rolls-Royce Dart engine, first launched in 1953.

Also amongst these was the Vickers Viscount aircraft, the first commercial aircraft to be powered by a Gas turbine engine (the Dart), which proved a very safe and reliable engine and aircraft, popular with both passengers and airlines alike. Throughout the 1950's and 1960's, gas turbine power replaced piston engines in major commercial and military operations. Rolls-Royce won a significant share of world orders in the market in both fields and by the mid-sixties had an acknowledged position as a major manufacturer of both military and commercial aviation.

Sir Denning played a major role in all of these achievements as well as being chiefly responsible for obtaining the business of supplying power units for the new British nuclear-powered submarine programme for Rolls-Royce. He was instrumental in forming the subsidiary company, Rolls-Royce & Associates, looking after the company's fledgling nuclear business. The US Admiral H.G. Rickover, a significant figure in the US nuclear powered submarine programme refused to work with any but Rolls-Royce and the mutual respect and regard had between the two men played an important part in this. Sir Denning became a director of Rolls-Royce Ltd. in 1950 and, in 1954, Managing Director of its Aero Engine Division, then later Chief Executive of Rolls-Royce and finally its Chairman in 1969.

Sir Denning regarded the liquidation of the company as a failure and a personal failure given that he was so closely involved.

Sir Denning was widely read and a man of high integrity. He was interested in people and their motivation as both individuals and groups. He was President of the Society of British Aircraft constructors and a member of the National Economic Development Council from 1964-67. He was a member of the Universities/Industry Joint Committee of the Confederation of British Industry, a governor of the London Graduate School of Business Studies, and a member of the council of Manchester Business School. He received honorary doctorates from the universities of Brunswick, Nottingham, and Wales, and was a Fellow of Imperial College London (1968). He was knighted in 1963.

He joined the Institution of Mechanical Engineers in 1947 and was elected an Honorary Fellow in 1969. Despite the insolvency, it has to be said, he continued to contribute greatly to both Rolls-Royce and to the engineering industry in the UK.

A CAREER AND PERSONAL PROFILE OF SIR DAVID HUDDIE (Rolls-Royce Limited, Main Board Director, 1960–1970, and Managing Director, 1965–1970).

Information extracted from 'The Dictionary of Irish Biography' as contributed by Stephen Bragg, who worked at Rolls-Royce in developing the Blue Streak missile but had left the company three days before it became insolvent in 1971, to take up the Vice- Chair of Brunel University. Additional quotes taken from the Obituary in 'The Independent' Newspaper.

> Sir David (Patrick) Huddie (1916–98), engineer, was born 12 March 1916 in Cavan, Co. Cavan, sixth among four sons and three daughters of James Huddie, farmer, and Catherine Huddie (née Frizelle), of Clara House, Lisnaskea, Co. Fermanagh. He was educated at Sligo Grammar School and Mountjoy School, Dublin, before entering Trinity College Dublin (TCD). He graduated with first-class honours in mathematics and economics, being awarded gold medals for both subjects. After leaving TCD, Sir David taught for a short time in Edinburgh, and then joined the

British army at the outbreak of war. He was soon drafted to Rolls-Royce Ltd, the company where he spent the next three decades working on engines for aircraft. 'DPH' (as he was always known at Rolls-Royce) first directed his mathematical skills to the calculation of stresses in the moving parts of piston engines, particularly the Merlin engine which powered so many Royal Air Force planes such as the Spitfire and Hurricane fighters, the Mosquito fighter/bomber, and the heavy Lancaster bomber, as well as the American P-51 Mustang fighter.

> "Huddie then became closely involved with the conversion of the Merlin for airliner use ("our baptism of fire") learning that the thousands of cruising hours expected in civil air transport was as demanding for an aero engine, in a different way, as the shorter bursts of violent activity in war". (*Obituary by Andrew Nahum, The Independent, 16 June 1998)*

He was promoted to assistant chief designer in 1946, but two years later moved over to engine development work. DPH was chief development engineer from 1953 and subsequently assistant chief engineer (1956) and was responsible for the development of the outstandingly successful Dart engine, which was installed in the Viscount, the world's first turboprop aircraft, one of the most successful British civil airliners in the whole post-war era.

By the 1960s DPH had moved up to more managerial positions, first on the commercial side as general manager, Sales, and Service (1958); commercial director (1959); general manager, Aero Engine Division (1962); and then, in 1965, managing director of the Aero Engine Division. He had been elected to the main board of the company in 1960. DPH worked hard, and inspired others to work with him, he was a good judge of talent. Sir David resigned on grounds of ill-health late in 1970 and severed his connections with Rolls-Royce for a long time.

> "In the aftermath of the bankruptcy Sir David Huddie behaved with immense restraint and honour. Leaving Rolls-Royce for Imperial College he and Sir Denning Pearson effectively shouldered the blame, accepting that the important thing was the success of Rolls-Royce, and never indulged in any self-justification, even when the engine and its successors were technically and commercially vindicated. However, it is good that, in the years before his death, he was brought back into contact with the company by well-wishers. He said of the engine last year that "as a failure, it hasn't done badly"." Ibid *Andrew Nahum, Independent, 16 June 1998.*

When his health restored, he was appointed a senior research fellow at the Imperial College of Science and Technology (ICST) in London.

Alan Swanson, FREng. Head of Mechanical Engineering 1978-1983, Pro Rector (Educational Quality) ICST, wrote in the *Independent* Newspaper on the 6th August 1978, to expand the above information on Sir David, explaining that he had . . . *'been invited to join the Mechanical Engineering Department at Imperial by its then Head, Professor Sir Hugh Ford, who was developing his ideas about engineering education and in particular was seeking a better integration of university education and industrial training, to produce graduates who would be as sound in engineering science as their predecessors but better able to apply it in the real world'*.

As well as assisting in devising the 1-3-1 Total Technology Course, (three years in university sandwiched by two years in industry), Sir David was able to utilise his considerable experience in enlisting industrial support. He did this so successfully that most major employers of engineering graduates were partners in the Total Technology course preparing engineers for management roles. The first students started the Imperial College part of their course in 1975. He would later also devise the same for the post graduate students on a 1-4-1-*x* ('apparently '*x*' being due to the post graduate part being vague') and largely from the same pool of sponsors. His work left Imperial with an array of sponsors for both the three and four year courses that were the envy of others. Many young engineers would have benefited from Sir David's work in these programmes.

Sir David retired in 1980 but was elected Honorary Fellow of ICST in 1981. He was also elected to the fellowship of the (Royal) Academy of Engineering in 1981.

Reflections on Sir David Huddie's lifetime career and contribution to Rolls-Royce Ltd. by his son Bob Huddie.

> *"My father had been promoted rapidly through the ranks of Rolls-Royce which meant that the challenges of firstly selling and secondly delivering the RB211 dwarfed anything he'd taken on before, just as they dwarfed anything Rolls-Royce had taken on before.*
>
> *I have no doubt, having met many of the other senior people in Rolls-Royce by that time, that he was the right man to lead the aero engine division and the right man to lead the company into this huge endeavour and no one could have worked harder at it than he did. I doubt whether the contract could have been obtained at all without*

the huge personal investment he made in it (going to live in New York for the necessary time was only a part of it). However, if both he and Rolls-Royce had experienced anything as big as this before, things might have gone better and perhaps the crisis of 1971 could have been avoided. In the end, the stresses on the company and on my father proved too much for them. It broke the company and broke my father's health and career. Companies are managed differently now, and critics can point out some bad decisions and failures of execution, but in a hugely complex undertaking, that's par for the course.

Looking back, what I saw of the RB211 crisis was through the lens of my family and my father's role in that, as husband and father. The first thing that was crystal clear at home was how deeply he loved the company. Early in World War II, he was plucked from the groves of academia and ordered to report to a satellite Rolls-Royce engineering office in Belper and of this he later said simply, "I knew I had come home". Rolls-Royce's engineering group was full of clever and able people, and the collegiate atmosphere and technical camaraderie were meat and drink to him. His love of challenges, focused hard work, articulate and open personality and drive duly led to promotion within the engineering function and later into general management, culminating in his appointment as Managing Director of the Aero Engine Division before the crisis of this book took place. At home we saw all this through an increase in our standard of living (not my father's main incentive), through the chauffeur-driven Bentleys that eventually replaced car-pooling to take him to his office in Derby and back, through the young graduate P.A.s who worked for him and came often to our house, and through the inexorable increase in his workload, most obvious to us in telephone time spent with colleagues at all hours. We met many of his colleagues at home, from higher up the organisation chart as time went on, whom to a man, shared levels of love of, pride in, and dedication to Rolls-Royce, which over the course of my life I have rarely seen matched by employees of other companies. For many of its people, at all levels, Rolls-Royce was their lifetime employer or close to that. It wasn't a company people wanted to leave. It's hard to explain this feeling of belonging simply in terms of the obvious excellence and reputation of its engineering but working for Rolls-Royce gave an extra dimension to the lives of many of its employees in those days.

At home, he encouraged us to express our opinions and schooled us informally in the art of debate. His ability to explain things and convey his enthusiasm for them was exceptional and perhaps it was this that also underpinned his ability to inspire devotion in younger people in the company. I well recall Robin Forrest, one of a succession of able, young Rolls-Royce graduate apprentices, who was my father's P.A. for a while, saying that "If Sir David had asked me to jump off the edge of the nearest cliff, I would have seriously considered it". Amongst the other Rolls-Royce executives who worked closely with my father was Richard Turner, later to use his unrivalled rapport with airlines and other customers worldwide as Rolls-Royce's Marketing Director, and Ralph (later Sir Ralph) Robins, who was later to lead Rolls-Royce most ably as Chairman for many years (though my brother Patrick and I knew him better through Belper Meadows Cricket Club). It was a productive stable.

One of his biggest fans and ablest counsellors was a hugely talented but down-to-earth American lawyer called Phil Gilbert (Gilbert, Seegall and Young, Rolls-Royce's US appointed lawyers), whom I came to know well (his lovely house in Westchester became my home-from-home when I spent a year at Harvard) and with whom I later worked on several business matters of my own. There's a good story about my father and a colleague in New York, searching for a US lawyer for Rolls-Royce. Phil was the last one they interviewed and, in contrast to the other candidates, when asked what he knew about US aviation and airlines, he replied "Nuttin". This disarming frankness got him the job and he turned out to be a huge asset to Rolls-Royce in the USA particularly during the stressful episodes dealt with in this book and whilst serving as President of Rolls-Royce Inc. Phil and I often talked of Rolls-Royce, and he was able to convey to me how my father's openness and 'can-do' attitude, perhaps unsurprisingly, went down so well with US customers.

My father's main, chosen relaxation from Rolls-Royce work was more work, though of another nature. The part of the weekends that he spent at home after, in those days, Saturday mornings in the office and later only when international travel allowed it, were always devoted to toiling in his garden, which despite its physical content he considered less arduous. He wanted to work and see its results at all times.

Rolls-Royce's decision to break into the most competitive aero engine market in the world with a high-thrust, innovatively engineered engine for wide-bodied aircraft was undeniably audacious and all those involved, my father included, knew that very well. They also knew that failure to achieve it was likely to limit Rolls-Royce to a relatively minor part of the rapidly expanding commercial aero engine market. My father's dedication to the success of the RB211 couldn't be maintained without some of the stresses that came with it becoming evident at home. His promotions had been rapid and perhaps a more cynical, more widely experienced and stress-toughened executive would have known better than my father that worrying about the things you can't control doesn't help a lot.

Could harder bargains have been driven against formidable competition? Maybe. Could some adventurous engineering options have been pruned earlier? Maybe. But maybes they are and maybes they remain. In the end, Rolls-Royce couldn't deliver the RB211 engine in time to meet the contract delivery dates with Lockheed. When the contract was signed in 1968, it had put my father's and Sir Denning Pearson's pictures on the front pages of national dailies and earned my father a knighthood to match Sir Denning's in 1963.

Now that same contract brought them both down.

An aero engine's life, from initial outline conception until the last sale of a spare part, can span half a century or more, so companies like Rolls-Royce are best judged in a long-term context. The RB211 crises were finally overcome, and its unique three-shaft architecture lives on in Rolls-Royce's world-beating family of high-thrust Trent turbofans, for almost 30 years now, the mainstay of the company's business.

I hope my father and Sir Denning Pearson can look down from the 'great drawing office in the sky' and take some satisfaction in that.".

A CAREER AND PERSONAL PROFILE OF DANIEL J. HAUGHTON (Lockheed Aircraft Corporation, Chairman, 1967–1976)

The following is supplied by courtesy of Sherm Mullin, Lockheed. (The Star Dusters organisation for retired or former employees) and from the National Aerospace hall of fame archive (USA).

"Daniel Jeremiah Haughton was employed by Lockheed Aircraft Corporation from 1939 to 1976. A towering workaholic, for 20 years, from 1956 to 1976, he ran Lockheed with tight control and intense attention to detail, often calling 6am meetings with his executives. His commitment was unmatched by any other Lockheed executive in its history and his devotion and commitment was total.

Born in Alabama on September 7th, 1911, he graduated from the University of Alabama in 1933 with a B.Sc. degree in business administration. After working for several companies, he joined Lockheed in 1939 as a systems analyst in the manufacturing branch at Burbank.

During World War II, he analysed aircraft manufacturing operations in detail and created a manufacturing control system to establish budgeting, cost control, and schedule, reporting the detailed level for each department. This was established as a company policy. His impressive hard work and results brought him to the attention of Vice President Courtlandt Gross, younger brother of CEO Robert Gross. Haughton's career then advanced rapidly.

When the Lockheed Georgia Division was established in 1951, Haughton moved to Marietta as Assistant General Manager, soon becoming Vice President and General Manager. Division performance was excellent. In 1955, Robert Gross made the decision that the C-130 transport aircraft, designed and prototyped in Burbank, would be produced in Georgia. Haughton launched the effort very successfully. Robert Gross stepped down as Lockheed President in 1956 and was succeeded by his brother Courtlandt. Dan Haughton was promoted to Executive Vice President, with full authority over all Lockheed divisions. When Robert Gross died in 1961, Courtlandt became Chairman and CEO. Haughton became President. In 1967, Courtlandt Gross retired, and Haughton moved up to be Chairman and CEO.

Dan Haughton's highest priority as CEO was to initiate a new commercial airliner programme after Lockheed California Company had lost out to Boeing in the supersonic transport (SST) competition. This led to the rapid design of the L-1011 TriStar wide-body aircraft, powered by three Rolls-Royce RB211 jet engines, which competed directly for airline customers with the Douglas DC-10, with three GE engines. Note that in 1967, facing major financial problems with its DC-8 and DC-9 programmes, Douglas was acquired by McDonnell, forming McDonnell Douglas.

Haughton personally led the L-1011 marketing campaign, with initial focus on American Airlines. In February 1968, American selected the Douglas DC-10. Haughton responded by making aggressive fixed price offers to others. TWA and Eastern promptly ordered TriStars. They were followed by Delta and Air Holdings, bringing total orders to 144, plus options for more. Production was started and a new assembly plant at Palmdale was rushed to completion. The TriStar made its first flight at Palmdale in November 1970."

Daniel Jeremiah Haughton began his career with Lockheed Aircraft Corporation as a systems analyst in 1939 and rose to become Chairman 28 years later. Within two years of accepting a position with Lockheed he was made an assistant to the Vice President of Vega Aircraft Corporation, then a subsidiary of Lockheed. He held that post until he became works manager in 1943. Later that same year, Vega merged with the parent Lockheed company.

From 1949-1951, Haughton served as President of two Lockheed subsidiaries, Airequipment Company and Aerol Company Inc. In 1951 he became Assistant General Manager of Lockheed's Georgia Division in Marietta, Georgia, and the following year he was elected Vice President and General Manager.

Under Dan Haughton's leadership, the Georgia division began the development of what was to become one of Lockheed Aircraft's brightest stars, the rugged and versatile C-130 Hercules Turboprop powered, troop and cargo transporter.

He was elected Executive Vice President of the Corporation in 1956 with authority over its three divisions and four subsidiaries, and a director in 1958.

In September 1961, he was elected President of the Corporation. At that time Lockheed was the 28th largest corporation in the US with a half billion dollars in total assets and 70,000 employees worldwide.

Dan Haughton was named 'management man of the year' in 1966 by the National Management Association, for his 'dedication to the concept of managerial teamwork in advancing company aims and for application of his highly ethical personal code to business affairs'. He was said to start work at 5 am each day and employed three secretaries.

He was elected Chairman and Chief Executive Officer of the Board on May 2nd, 1967.

Haughton was National chairman of the US treasury department's industrial payroll savings bond campaign that same year. The Lockheed board chairman received the marketing executive of the year award in 1968 from 'Sales and marketing executives international' for his direction of the successful marketing campaign programme for the Lockheed L-1011 TriStar airliner and received the National Aviation Club 'award of achievement' in 1969.

Dan Haughton devoted a great deal of his time to public service activities, locally and nationally. He served as Chairman of the board of the National Multiple Sclerosis society and served as a member of the board of trustees of the National Security Industrial Association, the National Space Club Foundation, and the Boys Club Foundation of Southern California.

From 1969 to 1975 he was a member of the board of governors of the American National Red Cross. He retired as Lockheed's chairman in February 1976. ***The National Aerospace Hall Of Fame Archive (USA)***

Sir Ralph Robins was quoted as saying 'He taught us a great deal'. Describing his manner, by a friend in the US, recorded in Peter Pugh's book, Ibid , as 'red-dirt southern courtliness – nothing affected, nothing high-flown and yet unmistakably gentry'. He was described as being deeply and personally hurt by the decision by Rolls-Royce to declare bankruptcy and could not understand the action the British government had taken in apparently "repudiating" a potential £1,000–million contract to supply jet engines for Lockheed's L–1011 jumbo jet.This may go some way to explain his diplomatic and restrained reaction to the bankruptcy quoted in Ibid., *(Hayward K. (1983), pg115-116)*; 'Flight' magazine article, *11th February 1971 p.182,* as:

> *"We were completely surprised and appalled at the precipitant decisions made by the Rolls-Royce Board and the sudden withdrawal of the British Government's financial support for this key industrial firm. The Government's stand today represents a distinct change from its announcement last November . . .* [12]*. We deeply regret that this great British company with its long and proud reputation should be brought suddenly into such a desperate position.'.*

Sir Denning Pearson certainly believed that Dan Haughton formed the impression that the Government's commitment to Rolls-Royce went beyond the formal provisions of a launch aid agreement. (Ibid., *(Hayward K.) pg101.*

12 The announcement referred to was in relation to HMG making public its willingness to provide Rolls-Royce with additional launch aid of 70% of the increased costs over £75 million, of the RB211-22 to a maximum of £42 million, subject to a review of the figures supplied by Roll-Royce by independent accountants (Cooper Bros. & Co); satisfactory contractual arrangements and a limitation for a period, of any distribution on the company's ordinary share capital, making a total contribution by HMG of £89 million.

SOME EARLY RB211 TECHNICAL CHALLENGES ALONG THE WAY.

The following contributions provide an insight into some of the major technical problems which faced the RB211 engine designers in the early design and development stages of the RB211 engine programme.

HYFIL – 'Game Changer or Game Breaker' Commentaries By John Coplin, Bill Griffith, & Alan Buxton

JOHN COPLIN (former Rolls-Royce RB211 Chief Designer and Director of Technology and Design)

HYFIL AND COMPOSITE COMPONENTS AND THEIR EFFECTS ON RB211 ENGINE DESIGN

"I have vivid memories of those early days of the RB211 and the delicate path that had to be found, within the most unreasonable technical design constraints. However what David Huddie, as leader of the RB211 Sales campaign, had promised had to be honoured.

It was based on the former Director of Engineering, Aero Engine Division, Adrian Lombard's dream, some of which flouted the rules of Mother Nature. The only way forward was to find the lightest structure that the laws of physics allowed. Lombard was totally committed to introducing composite materials for the high-duty components at the front end of the RB211 engine and saw the Rolls-Royce military lift engine work on the US/UK XJ-99 lift jet programme, as the means to that end. Sadly, he had already passed away by the time it became clear that carbon composites dictate design requiring different structural concepts from those with which the company was familiar.

I had worked for Lombard as his technical assistant and his understanding of the physics appeared to be less than what was needed to achieve a satisfactory aero engine design, especially for high-duty components. I believe he may have had no detailed knowledge of the US/UK, XJ-99 lift jet programme (for which I was chief designer) we ran in the mid-sixties, as security was very tight indeed, with precious few personnel having access to the research. It was this programme that revealed the constraints imposed on engine design, especially if using carbon fibre and the best epoxy resins. I was therefore a natural choice for programme designer for the RB211 and RB207 engines. I was appointed in 1968 and took up that post in May of that year.

Lombard was an extremely assertive engineer and revered by us all for his outstanding contributions. He was a dominant force and Rolls-Royce 'seniors' trusted his judgements, which were often delivered with very high confidence. You can design successfully with composites, but you must respect all strengths and all weaknesses, and those who laid down the basics of the RB211 and RB207 engine architecture skated over the weaknesses.

These had been exposed by my XJ-99 design work. The lift Jet programme was rigorously tightly controlled by US rules so the RB211 teams were completely unaware of what had been exposed. Equally the lift jet team had no sight of what was being offered up for the high bypass ratio/high pressure ratio engine for long life commercial aviation duties. Lombard was simply not aware of the rules that must be adhered to when designing with resin-based composites,

The early design team lacked this knowledge and continued to design with rules suited to homogenous metals, not materials strong in one direction only. Moreover, the Laboratory stuck rigidly to forged HP turbine blades when we had already learnt how to make cast blades in better materials with better cooling. The XJ-99 Turbine blades were far ahead of anything that could be achieved with forged materials. Once the Lab understood this, they showed top of world class skills, first with better freezing techniques followed by single crystal with controlled crystal orientation. They became world leaders driven by a deep understanding of the physics pertaining to cast metal alloys.

David Huddie, like Lombard, was a brilliant man and was totally dedicated to winning an engine commitment in the USA from either Douglas or Lockheed. Perhaps, because of his own doubts about the giant step change in design from the Conway to the RB211, he committed to a 'stepped development programme' to assist this. This helped him to persuade the prospective buyers of the RB211 in the USA, but it escalated the design and development work required enormously, to the point where the design force had to be increased massively. This led to the use of designers spread far and wide.

The few with expertise, were supervising designers with less expertise, often involving lost time in travelling. At the early meetings chaired by Huddie every Monday morning, I had to deliver news about how we were having to find ways of achieving most of the benefits of composites by methods that could be underwritten with some confidence. These changes were understood, as the evidence in support was convincing, but it was hard for David Huddie, who was excellent in understanding and in backing what needed to be done. To make things worse, Lockheed persuaded Rolls-Royce that they needed to take thrust at the front of the engine on the fan case and the compressor aerodynamic designers discovered that the intermediate pressure (IP) compressor required variable inlet guide vanes (VIGV). Together, these changes totally destroyed the main composite structure, forcing a change to steel.

In the early stages of the RB211 sales campaign, I escorted David Huddie as he visited Lockheed and the airlines to explain how we were progressing, and I had to do many of the presentations. I was always backed strongly by Ray Thompson from Rolls-Royce manufacturing who was upset by the implications for production that these changes would make and that these were severe!

Some 50 plus years have now passed, and Rolls-Royce still has no wide-body jet engine with a composite front-end in service, so I have no regrets about my early decision for the removal of many high-duty composite components. There were, of course, other issues affecting the RB211 at that time, such as the high pressure (HP) turbine blade, and the complexity of system integration with the aircraft. At one of the said Monday meetings, I announced a decision to increase the gap behind the HP compressor to give a cleaner entry of the high-pressure air into the combustor. We needed a consistent flow with the lowest pressure loss. We had the opportunity to do this when Lockheed wanted some changes to the engine installation, and we could embrace this within the package of their required changes.

At a time when a single A4 page drawing, for example, could take 30 minutes to send via the latest Fax machine, frequent meetings with Lockheed would take place, held mainly in the smart temporary buildings on the Burbank airport car park. It is also interesting to point out that one of Lockheed's full-time representatives in England took up semi-permanent residence in Adrian Lombard's house, in Derby.

The XJ-99 programme work exposed what would not work, but it also showed what would. And the XJ-99 did meet its objectives, as the highest thrust-to-weight ratio for a jet engine. It also had the highest firing temperature of any engine at that time despite having a much more difficult turbine to design. Evidence from the XJ-99 programme was used to convince David Huddie that we had to design within the rules of physics and again this was done at the regular Monday morning meetings in front of all interested parties. I had the highest regard for Lombard and Huddie and had those who preceded me designed for composites we could have had a composite front end to the RB211. The design offered was fatally flawed but the constraints were set, and these constraints were immediately tightened straight after the contract was signed."

BILL GRIFFITH

My Hyfil Experiences

"As a technician apprentice, I worked at Hucknall on Hyfil fibre development between July and August 1969. There was already a production carbon fibre plant in operation and as a potential development to enhance fibre mechanical properties, the methodology was to use a fluidised bed of glass beads to transfer heat to the Rayon fibre instead of using an air furnace approach.

The results were good, but no attempt to scale up was made, perhaps for good practical reasons. The production plant could produce metres of carbon fibre daily compared with a volume from the fluidised bed that was more like a hair extension in length and width.

Nearing the completion of my apprenticeship in May–October 1971, attached to the Main Works Process Laboratory, I assisted Brian Forman and Joe Strong in the nickel plating of Hyfil Conway blades as an attempt to prevent the erosion and delamination experienced during testing. Sadly, the technique was a failure, the nickel plating readily peeling off during spin testing. I had a scrap electroplated Conway blade for many years but having lugged it from site to site as my job changed, I eventually disposed of it in one of the many office clear-outs, something I now repent at leisure!

An alternative erosion coating, I think made by Desoto Titanine based on either polyurethane or rubber, also failed to prevent delamination. I am a little sad General Electric resolved these issues including the use of a titanium leading edge strip that was also used in development at Rolls-Royce and I often wonder that if someone had consulted the fabric weaving industry a solution to binding the lay-up components together could have been found. However, the fallout from 4th February,1971 caught up and put paid to any further development. Perhaps worst of all, the Hyfil plant was disposed of and the composites technology all but thrown away until its resurrection in recent years with the GKN/RR CTAL composite fan programme – also using an anti-erosion coating based on polyurethane resins.

My birthday being February 5th, so close to the receivership date, it was a pretty poor birthday present but at least I survived and completed 41 years at Rolls-Royce before taking early retirement."

ALAN BUXTON

RB211 Composites and Alfreton Road Site

"I moved to the Composite Materials Department at the Rolls-Royce Alfreton Road site in Derby in May 1968 (about eight months after Rolls-Royce took over the site) to work in the Production Engineering office. A considerable amount of the manufacturing capacity was taken up on parts for the RB211-06 and -22 development engine programmes: namely Hyfil fan blades, front bearing housings, and IP compressor casings. Also, Hyfil Conway LP 1 (low pressure) blades which were being flight tested by BOAC (now British Airways) on the London to Johannesburg route.

The first RB211 fan blades were of a straight double-dovetail root design. The curved single dovetail root followed on and engineering changes came through thick and fast during 1969 and 1970, attempting to resolve problems. In late 1970 and early 1971 you could tell that the pressures were building in engineering and major financial restraint was being felt across the company.

On Thursday 4th February 1971, I arrived at the Alfreton Road site as usual. At about 10.30am a colleague in the office (Fred Allsopp) was on the 'phone to Bristol Aeroplane Plastics, which was going to be the production facility for composites materials. At 10.50am, Fred came off the phone, saying that all the employees at Bristol had been called to the canteen for an announcement.

Work carried on as normal at Alfreton Road; from where I sat, I could see the site gates still open and vehicles coming and going as usual. At about 11.20am, Fred received a call back from his colleague at Bristol, saying that "Rolls-Royce had gone bust and into receivership at 11.00am". Another colleague had to contact a subcontractor at about this time, only to be told that "we had gone bust" and they were not sure whether they should even be talking to us!

It appears that Alfreton Road was forgotten about until shortly before midday. At 11.00am when the receiver officially took over, all the company gates were closed and locked to prevent goods coming and going. The main company phone switchboard was also shut down. As we had an independent switchboard, several of us were able to make external phone calls to wives and family with the news before the official announcement. One story I heard was that a train delivering fuel for the test beds on Sinfin A site was locked in for several days during this shut down.

Some days after the 4th, the receiver gave permission for an engine test, incorporating a collection of simple modifications to prove that it could attain the specified requirements. Alfreton Road was asked to produce some wooden trailing edges to be attached to the rear of the A-frame structures in the rear fan case, to smooth out the air flow. (The development drawing specified 'straight grained mahogany'!) These were bonded on and wrapped with a layer of glass fibre. Further to this modification, during engine build, all the gaps between the IP compressor fairings were filled with a composite filler to help reduce drag and smooth out the air flow from the fan, the overall aim being to increase the compression.

The resulting engine test proved successful in demonstrating that the RB211 engine could achieve the specified thrust and was a viable marketable product."

DR. IAN MACPHERSON: Some Early RB211 Technical & Miscellaneous Development Issues.

1. *Component Department*

"Previous accounts of the development of the RB211 engine can be found in the James Clayton lecture by Eltis and Wilde (Eltis, E.M. and Wilde, G.l. [1974] The Rolls-Royce RB211 Turbofan Engine, Proc. Inst. Mech. Eng., 188[1], 549-575) and the 31st Short Brothers Commemorative Lecture at the Royal Aeronautical Society by Phil Ruffles (Ruffles, P. [1992]. The RB211: The First 25 Years. Royal Aeronautical Society, N94-23570).

In the latter's major account of the development of the RB211 engine in his book 'The History of the Rolls-Royce RB211 Turbofan Engine' *there is some mention made of the contributions of the component engineering departments which had to evolve and implement the technical advances needed. In the very early stages, these were hampered by the inadequate provision of test rigs. My responsibilities covered the combustion blade cooling and the turbine departments from 1967 to 1971 and my colleague Roy Hetherington, Assistant Chief of Advanced Engineering, covered fan and compressor design.*

2. ***Combustion Department***

The manager was Terry Durrant, ably assisted by Dick Carlisle, Fred Murray, Arthur Lefebvre, Tony Wassel and others. The development of such a large annular chamber took the previously demonstrated technology beyond the capability to meet the needs.

2.1 ***Major Issues the Team had to Deal With***

There was not a large enough high-pressure rig to run a full-scale chamber at rated power. The essential feature of temperature traverses at full power did not become available until well into the late 1960s. Traverses required pinpoint location of racks of thermocouples that could be stepped around by about 5^0 angularly to establish the uniformity or not of the outlet temperature pattern. This took several hours and required overnight runs because of the electrical power demand of the reciprocating Belliss and Morcom compressors that fed air to the rig. Hotspots at the exit to the chamber placed an unacceptable level of stress on the HP Turbine Nozzle Guide Vanes (NGVs), leading to premature cracking of their leading edges and vane failure. Attempts made to examine the burning pattern in the chamber through the 'spy hole' at the end of the rig detuner were desperate and fruitless.

2.2 ***Internal Flow Patterns***

Studying the entry of the airflow into the chamber could only be done by using a water analogy rig in which a single sector of the annulus, made in Perspex, was fed water infused with air bubbles to trace the primary zone recirculation vortex and dilution air patterns, as recorded on the movie camera. This could not reveal anomalies in the overall pattern in such a large annular chamber with little constraint on erratic internal flow patterns. There was no capability to model the gas dynamics of internal flows.

2.3 *Wall Cooling*

The first runs on the full annular chamber, at relatively modest temperatures, were disastrous. The chamber emerged as a badly 'distorted dustbin' with holes burnt through and major hotspots in the walls, shown by thermal paints. The chambers were fabricated at the sheet metal works at the Rolls-Royce Mountsorrel site. The prior technique of using rings of 'wiggle strips' to join successive elements of the chamber (previously successful with the Dart, Avon and Conway chambers) was intended to introduce a thin film of cooling air along the inner surface of the chamber walls.

A wiggle strip was relatively easily made by crimping a corrugated pattern into a ring of Nimonic 75 that matched the diameter of the chamber. They looked like extended pastry cutters. These strips had to be abandoned in favour of rings, about five spaced along the length of the chamber, with drilled holes (2mm at 5mm spacing, about 2000 per chamber) so-called 'top hat' holes, but at much higher cost. These proved successful in the end after much trial and error, thermal paints showing uniformly cooled regions of casing.

2.4. *Carbon Build Up on Fuel Injectors*

These were early fuel injectors with air-blast atomisers invented by Arthur Lefebvre, each with a burner shroud. Eighteen were located in the back plate in the chamber, fed by HP compressor air. Late in the engine development programme at simulated high altitude cruise conditions, carbon built up on each injector and burner shroud. This became hard carbon and had two serious outcomes. The distorted fuel injection pattern was reflected in the adverse temperature traverse at the entry to the HP turbine. More importantly, the hard carbon was judged to adversely affect the relight capability after a flame-out in an aircraft at cruising altitude and became a potential 'job stopper' for the RB211 programme. Fortunately, some careful analysis of the airflow to the burner shroud showed how a modest increase in fresh air to the shroud resolved the issue.

2.5. *White Smoke on Cold Start*

Starting an engine on an L-1011 aircraft, early morning, that had been parked overnight in sub-zero temperatures (e.g., as witnessed by myself at Toronto Airport in winter) produced immense amounts of white smoke, un-vaporised fuel, that almost obscured the aircraft.

This was eventually cured by progressive development of the ignition and initial fuel injection scheme but not before significant adverse publicity.

3. ***Blade Cooling***

The manager of this area was Gordon Halls.

It was clear after the Lockheed contract had been signed that the existing capability to provide cooling to the HP Nozzle Guide Vanes (NGVs) and Stage 1 turbine blades would be inadequate. (See my later remarks about David Huddie's negotiations.)

Manufacture of NGVs by the lost-wax casting process was first advanced by using a ceramic core that generated a large internal space within the vane with a central tube that could be fed with HP compressor air. Running a rack of a few blades on a modest high temperature static rig and using thermal paints to reduce surface metal temperatures gave an indication of problem areas. This level of design included a few internal features to promote good heat transfer to the cooling and was still susceptible to compressive creep and failure by cracking at the leading edge. I made a fruitless effort to introduce internal ribbing in a test vane by using a dental drill to cut small grooves in the surrounding of the ceramic inner core before casting.

The first attempt at cooling the original forged RB211 HP turbine blade was by using arrays of fine holes into a central chamber which were also fed by HP compressor air that had bypassed the combustion chamber, flowing into a cavity at the base of the blade. A significant loss in the pressure of this 'cool air' as it bypassed the inner wall of the combustion chamber prevented the use of such film cooling holes at the leading edge (stagnation point) limiting the turbine entry temperature. The first technique for 'drilling' the fine holes used the electro jet technique invented by Dr. Ken Payne, working at the Rolls-Royce Advanced Research Laboratory at Littleover. This technique was later to be superseded by 'spark machining'.

[Ian's son, Tim, who would have been aged about 11yrs at the time, recollects witnessing his Father's attempts to make the cooling channels in the Titanium compressor blades, held in a vice on his Father's work bench in the garage, using a dentist drill bit and witnessing his Father's weekend frustrations!]

4. *Single-crystal Turbine Blades*

Major developments came about in the late 1970s as the programme continued, leading to the multi-pass single-crystal turbine blade and NGV's fully, film cooled with air extracted from the compressor. Both turbine blades and NGV's in today's engines such as the Trent are single crystal and film cooled. The central materials laboratory and manufacturing played a crucial role in bringing this technology forward. The long-established intent to procure a dedicated blade-cooling demonstrator, the so-called High Temperature Demonstration Unit (HTDU) based on the core of an actual running Tyne engine, did not come into effect until well into the early 1970s.

The Turbine Department manager was Simon Byworth, assisted by Harry Tubbs and others. Turbine efficiency in the early RB211 development engines was poor, not matching the expectations from rig work. The problem was identified in the lack of consistent tip clearance of the blades in an operating engine because of differential axial growth in the carcass, due to a hot shaft and cold casing. The problem was overcome by mounting the static seal segment on the hot combustion inner casing rather than the cold outer casing.

5. *Fan and Compressors*

The early RB211 engines used the 'prior design processes' for setting the bend in the fan blade at root, mid-height and tip for optimum incidence for the air entering the fan. The blades also used short 'clappers' or 'snubbers' between the row of blades at mid-height to prevent contact by excessive flapping, introducing their own set of inefficiencies. The early RB211 exhibition engine still held in store at the Manchester Museum of Industry and Science is an example of fan blades with clappers.

Rolls-Royce used computers from the early 1960s and bought its first mainframe computer, an IBM360, in about 1967. At first the input was by punched cards, with the run taking a weekend and the output collected on Monday with even more punch cards, a process easily rendered fruitless by minor errors in the punching process. The later successful fluid dynamic calculations for the entire incoming flow around a fan blade that gave rise to the wide chord blade fitted to later RB211 and Trent engines, was driven through by my colleague Roy Hetherington for the original Hyfil blades. One has trivial memories of course, like the wide chord fan blade being exhibited some years ago as a 'standalone' item of elegant design at the annual Summer Exhibition at the Royal Academy in Piccadilly, London.

Miscellaneous RB211 Programme Issues

6. *Engineering Review Meetings*

These were held every two weeks. Before his death, Adrian Lombard chaired the event, and this function was later taken over by Ernest Eltis. Others attending included John Bush, John Coplin, Cyril Lovesey, Tom Metcalfe and Lindsay Dawson, with departmental managers. All current development problems and plans were for the component departments to solve, and these were major agenda items and were robustly challenged.

In particular two events come to mind:

By the late 1960s, it was becoming clear that the continuing RB211 development and production issues were severely taxing the financial stability of the company.

At one of the review meetings, Ernest Eltis commented that the Chairman, Sir Denning Pearson, had met with Anthony Wedgwood-Benn, the Minister for Technology and Industry under Harold Wilson's Labour government, and the financial future was assured and -

John Coplin recommending at a late stage, that the central core of the engine should be lengthened in an attempt to improve the pressure recovery and uniformity of the HP compressor output before the airflow reached the combustion chamber.

7. *David Huddie's Contract Negotiations with Lockheed*

The team that went out with David to achieve this was advised early on that offering excessive levels and thrust would put such factors as turbine intake temperature up to a level that could not be assured by present development standards. In those days there were no emails or indeed fax machines for communications which were sent by telex. These messages were viewed with growing concern as reports came back of the unavoidable escalating levels of performance needed to clinch the RB211 contract. Before the contract had been signed, Engineering had already placed orders with Eric Scarfe (Head of Production) for long lead-time parts for 10 RB211 development engines, at a cost of about £1 million each. These were judged to be incapable of meeting the agreed levels of performance committed in the final contract.

8. *Letter to Hugh Conway, Director, 12th March 1971*

As the receivership went ahead, Hugh Conway was appointed to take over the operational management of the company. The success of the RB211 development work in achieving contract levels of performance and type test durability on March 11th, 1971, seemed to be passing by any sense of support coming from the government. Roy Hetherington and I penned a letter to Hugh Conway (see following pages 95 & 96).

At least we were trying to ensure that the success was put in a prominent spot on the agenda of the Rolls-Royce senior management and the government."

FORMER ROLLS-ROYCE AND LOCKHEED EMPLOYEES' EXPERIENCES & RECOLLECTIONS OF THE RB211 PROGRAMME

BILL GRATTIDGE, Former Head of Engine Testing

My career involvement with the RB211 engine project between 1967–1983

"Following a reorganisation of the Test Department in 1968, I became Head of Engine Testing with responsibility for the Derby, Barnoldswick and Hucknall test departments. By this time the 'pre-rigging' concept had progressed somewhat, and a pre-rig shop had been built near to the RB211 test beds. This turned out to be an unexpected saving grace, because traditionally the experimental assembly shop was used to assemble engines, vertically, using pits, which enabled the engine to be raised or lowered as required. A little late in the day it was realised that there was insufficient headroom in the assembly shop for the RB211 to be assembled vertically and so a combined occupation of the pre-rig shop had to be arranged, so that the engine modules could be assembled horizontally in the pre-rig frame. The basic procedure was to attach the fan case to the pre-rig frame and then add the other modules, front and back, in turn. A major event was the first run of the RB211. The contract with Lockheed had stipulated that the first run was to be completed by August 31st, 1968, and the actual first run was achieved on August 30th, 1968, at about 1:30am in the morning, albeit with a slave fuel system. To make the first run of a new engine during the night had become almost a tradition, not necessarily deliberately arranged, but it did have the effect of reducing the number of spectators who, although understandably concerned, were not specifically required.

STRICTLY PERSONAL

internal memorandum AERO ENGINE DIVISION—Derby

from I. Macpherson
R. Hetherington

to H. G. Conway

our ref.: Mpn 3/CA

date 12 March 1971 tel. ML 9877
GPO 1406

As you are well aware many of the Senior and Middle Management of the D.E.D. are extremely concerned about the present negotiations on the future of the RB 211. Their concern is inevitably based on their limited knowledge of the situation but is nonetheless real. The attached note is the final version of a note that all of the Senior and Middle Management who know of its existence would like to send immediately to the Government. As you will see this note is tough in its content, which gives the measure of the intensity of opinion people have about this situation.

We have been asked to seek your guidance on the next step and would ask you to consider which of the following possibilities would be most helpful.

(1) The note could be forwarded through channels open to you, as an example of opinion, that could be useful in your own discussions.

(2) You may consider that the note could be most usefully sent by the Middle Management directly to the Government.

(3) You may consider that such a communication would be harmful however it is transmitted to the Government.

If your choice is the third possibility then that is the end of the matter as far as the people who have endorsed the note are concerned.

In this case, we think it would be of tremendous value in enhancing the relationships between Senior and Middle Management and the highest levels of the Company if you could meet informally with a group of ten to fifteen people to help us understand the apparent omissions in Government actions.

I. Macpherson
R. Hetherington

} Assistant Chiefs of Advanced Engineering, reporting to Lindsay Dawson

Internal memo to Hugh Conway

TELEX TO BE SENT TO LORD CARRINGTON
MR. CORFIELD AND BOARD OF ROLLS-ROYCE (1971) LTD

The Senior and Middle Management of the Derby Engine Division of Rolls-Royce believe the following points are of immediate relevance to the present position of the RB 211 project.

(1) Since 4th February the performance of the RB 211 has improved to a demonstrated hot day take-off thrust in excess of the 38,500 lb thrust envisaged for initial passenger carrying operation in the Tri-Star (Engine 10,011/2 run 11th March).

This performance was achieved without all the improvements to be built into the initial passenger carrying engines.

(2) Management steps to improve the efficiency of operations in the D.E.D. together with further cost reduction schemes for the RB 211 will give reductions in cost not envisaged on 4th February.

We believe these factors demonstrate that the RB 211 is worth enough Government action to ensure the continuation of the project in the Tri-Star.

In the present crucial situation we fear that not enough attention is being paid by the Government and Rolls-Royce (1971) Ltd to our customers of vital importance - the Airlines on which the whole project depends.

We have detected no signs that the Government is working with the Lockheed Company to show joint commitment to the continuation of the Tri-Star in such a way that the Airlines can have confidence in the intentions of the Government or Rolls-Royce (1971) Ltd.

The Government must realise that the aviation business calls for a unique type of relationship between engine and airframe suppliers working together to sell their joint products to the customers. It is quite inappropriate to treat the situation as one requiring bilateral negotiations between the Government and Lockheed in isolation. The current approaches are not consistent with the competitive business environment in the United States and in particular that associated with the Airlines. It is essential that Senior Representatives of the Government and the Management of Rolls-Royce (1971) Ltd., with full authority to negotiate financial terms should have immediate direct contact in the U.S.A. with the policy making levels of the Airlines.

Proposed telex to Lord Carrington

Chief Test Engineer

About this time, I wrote a paper to develop further the theme of the pre-rigging of engines, setting out to prove that investing in the equipment for this would save considerable capital expenditure in new test beds, as well as making it reasonable to remove an engine from the test bed in order to test the next available one, and thus improve test bed utilisation considerably, which was one of management's key criteria. This paper attracted some early attention and shortly afterwards I was appointed Chief Test Engineer RB211 by Ernest Eltis (then Director of Engineering) and Eric Scarfe, Director of Production a grand enough title but Programme Manager would've been more appropriate, with authority to set priorities across build specifications, assembly and test.

It could be said that this was the beginning of 'programme management' in Rolls-Royce, where people were given the responsibility and authority to coordinate the activities of many specialist groups in order to achieve an agreed programme. This was a learning curve for all concerned. The heads of the specialist functions had to cope with a programme manager interfering, so to speak, with their activities. The trade-off was that as a programme manager became more skillful and knowledgeable, he could make informed and constructive suggestions about priorities. Initially, I reported to Chris Webber, Chief Development Engineer RB211.

However, on February 4th, 1971, the very day after a very successful test to demonstrate considerable performance improvements with the RB211 engine, Rolls-Royce went into liquidation and one of the resulting changes was that Harry Malley, who was then General Manager at the Hucknall site, was transferred to the Sinfin sites with programme management responsibility for all experimental work.

Alex Fairweather and I suggested that we should report to Harry Malley because it seemed to be a better alignment with most of the areas with which we were involved. This was accepted. Following this, an RB211 control centre was set up at Sinfin, giving visibility of the status and target days for all engines in the programme. Thanks to my small staff on shifts, we also began publication of a daily status report on every RB211 development engine, including the estimated 'ready for test' date for engines on rebuild and a status comment for engines on test. This was colloquially known as 'The Daily Liar'. I forget what the size of the circulation was. We aimed to have a copy in all interested areas at the beginning of each workday. This, of course, was before the days of the internet or intranet. At that time, Lockheed key personnel were seconded to work with my staff to augment the visibility of the operation with mutual benefits.

Although many people were involved in all this activity, during those years of the 1960s and 1970s, due appreciation must be expressed to the energy and expertise of Harry Malley. Harry came to Rolls-Royce as a graduate from Liverpool University and worked mainly in the research areas of compressors and combustion. He was one of the characters of those days, clever with a common touch, never afraid to express his opinion to whatever level, whether it was the shop floor or to Board members. He was always full of enthusiasm.

Underneath his outward brusqueness there was more than a touch of compassion, as I knew personally through working so closely with him for many years. Once a month, the senior programme management team from Lockheed would fly over to Derby for a Sunday morning programme review. At Harry's insistence, this meeting was set for 11 o'clock on the grounds that, I quote, "We don't want to stop Bill going to church with his family and 'saying a prayer for us". Unfortunately, Harry died early, at about 56, I think.

Gradually the development of the RB211 engine succeeded and the engine and the three-shaft concept became the success it is today. In 1979 there was another reorganisation and the incoming management felt that my role of looking across build specifications, assembly and test had the effect of blurring the output performance of the individual areas. This was true, although I would claim that it was for the overall good, in that there was no point, it seems to me, of having a long queue of engines awaiting test and receiving a 'short of work' from the assembly line or vice versa. However, I was given the choice of doing programme management either for the assembly group, or the test group. I chose to move to the assembly group, reporting to Alan Jackson working alongside Eric Guilin, who was responsible for procuring the parts. During one of the regular fortnightly visits to the Barnoldswick site for programme meetings, Eric and I were involved in a road accident – someone else's careless driving. Eric never really recovered from this and eventually retired early.

At the time I retired in November 1983, the flight certification for the RB211–535E4 had just been received. In 1988, my wife and I went on holiday to Moscow. We flew British Airways in a Boeing 757 powered by RB211-535E4 engines. I managed to visit the flight deck after saying I was connected with developing these engines. I asked the pilot what he thought about the way the engines behaved. He said that about 18 months previously a colleague of his had had some in-flight fault with one engine, but it was a 535C and, of course, they have improved since then."

The 535C was the initial variant leading to the much-improved RB211-535E4 engine. (The RB211-535E4 engine is referred to in Chapter 10.)

BRIAN WOOD

The RB211 engine and some non-destructive testing (NDT) aspects

"As its name implies, non-destructive testing (NDT) is a means of assessing the integrity of the component under examination. I began working for the divisional NDT department under John Hislop in 1966, having moved within the company. We were then based in the Chapel building at Main Works. This curious name arose from the roof disguise used to try to protect what was then the Merlin test bed area from enemy bombing during World War II.

We later moved to the Metrology Lab. complex in Sinfin. I was involved in many aspects of the RB211 development during my time with the department before leaving to take up a new position, again in NDT, at Rolls-Royce & Associates on Raynesway, Derby, in 1973. My special technical input was the development and use of ultrasonic inspection methods for both manufacturing applications and on installed engines where operational faults had taken place. They were interesting days; the work was varied and new challenges for innovative use of this inspection tool were frequent. They included travel abroad for application and training of local teams for repeat inspections.

Here are some stories from those times.

One of the applications for ultrasonics was the rather boring testing of Hyfil test panels by measuring energy transfer across the panel in a grid pattern as an empirical evaluation of the effects of fibre layout and resin mix in trial blades. I don't believe it gave very useful information. Weaving was much more closely related to the use of this novel material than solid metals and I recall the Shirley Institute as one of those weaving research establishments which submitted samples for test.

The compressor drums for these engines were built up from individual titanium alloy discs, each electron-beam (e-b) welded together on the thin section flanges below the blade roots. This was a whole new field of assembly for the engine and Sciaky Inc. of the USA supplied a suitable e-b welding machine which was housed adjacent to us in the Chapel. The method is especially suitable for welding thin section joints in hard-to-reach areas such as those below the blade roots. Welding is done in a vacuum to prevent oxidation and on a drum-like component; the emerging beam from the weld must be absorbed in a matching alloy to prevent beam impingement on the underside of the joint on the far side of the drum.

Test drums had shown the success of the method and a batch of production drums was made. On inspection, these contained areas of contamination by a brittle material containing 40% iron and we were asked to find a method of sorting sheep from goats. The welding engineer told me later that the backing plate material had been changed by the bean counters to mild steel in an effort to save money, without any reference or consultation.

This rogue eutectic material, despite its high iron content, was non-magnetic. It had eddy current properties that could not be distinguished reliably from either of the parent materials and methods such as X-ray or ultrasonic testing stood no chance of giving a result. In desperation, we tried dye penetrant testing as the last resort and to our complete surprise the material rusted when left wet in air.

However, the department came in for heavy flack when all we could rather lamely offer as a sorting method was to leave the drums out in the weather and look later to see if they had rusted! I believe the whole batch had to be scrapped as the only safe course of action to take; it certainly was a rather expensive piece of penny pinching! Bird strike failures of the Hyfil fan blades put paid to their use and a solid titanium blade design was hurriedly put in place as the only alternative in the available time left.

Because of the titanium blades, size and weight, there was some concern that fatigue failure in the blend radii out from the root to the blade proper might take place and I was tasked with developing an ultrasonic inspection for the area once flying trials were taking place on the modified VC10 at Hucknall, with one pair of Conway's on the tail plane replaced by an RB211. Regular testing of the blade set took place to look for a developing problem.

I brought the programme to an abrupt halt on one of these inspection visits when I was not confident that what appeared to be an interference surface wave indication that could not be 'damped out' satisfactorily was not masking a more serious problem. Visual inspection of the local area was not easy because of the cramped viewing conditions but it showed nothing untoward. My colleague, who was being trained at the time, was equally uncertain. I woke up my boss John at 4am to ask for advice but all he could offer was to back my judgement; the consequences of a blade failure were too dreadful to contemplate.

My observations were checked, of course, and the flying trial was stopped for more detailed checking of this blade and others at Derby. I was barred from participation in those investigations and in the end no definitive cause emerged to explain the anomaly. However, the recommendation was made to Barnoldswick to polish the blend radii on future blades. Significant time had been lost in accumulating flying hours at a time when that was crucial for qualification.

The investigation into the RB211 fan disc failures was very well reported in the Rolls-Royce Journal, No. 8, which gave an excellent account of what happened.

What was missing from that story was our input. In the NDT Department we had been working on disc inspection following the second operational failure on 11th January 1973. Metrology Lab. thinking at the time was that there was an unusual form of sub-surface fatigue failure, peculiar to the IMI 685 alloy. It was even given the name 'low cycle fatigue' and this would lead ultimately to open surface radial cracking in the deep re-entrant shape of the disc at its thinnest section near the hub. After that, the life of the disc was perhaps one cycle.

Along with several others, I was sent out for testing in the USA right after the Everglades crash which was feared to have been caused by another disc failure, this time in the tail engine. The Rolls-Royce resident engineers had already visited the crash site where the accident turned out to be due to pilot error; all throttles were fully open to try and correct landing too early at Miami and all three engine discs were intact.

We were divided into two teams. One went on to the Lockheed Company in sunny California and I was sent to TRW Inc. in northern Ohio at Euclid, a suburb of Cleveland, then in the grip of winter. The main test method was by ultrasonic inspection, looking specifically for cracking in the orientation and position described by the Metrology Lab. The full inspection was supported by a hydrofluoric acid etch and special dye penetrant test in the same area. Each team also had a rather complicated manipulator for an eddy current examination.

Along with the others, we worked 10-hour days without breaks at weekends, and I was eventually rewarded with finding and marking up what turned out to be the only example of a fully buried sub-surface crack found at both testing sites. The disc went back to Derby for detailed examination of the section trepanned out of it.

On my return home after this rather exhausting tour I asked about my findings and was pleased to hear that the ultrasonic technique I had developed had found what the Metrology Lab. had predicted. They said it was only a few more operations away from failure. However, I was somewhat disappointed to find this news was now old hat and it was not mentioned any further. No glory there then!

It was just after this that I left to take up my new position at Rolls-Royce & Associates, which had been waiting patiently for my return, having made me the job offer while I was shivering in Euclid.

Some interesting non-technical information

TRW was then a major aero engine overhaul organisation in the USA, taking in work for GE and Pratt & Whitney. The factory was built in the 1940s and it is the largest single factory roof structure I have been into, measuring one-quarter of a mile in one direction and half a mile in the other. People got about inside on push bikes. One area closed off to us was the torpedo section; it was the only space to which we were denied access. The waste of power was truly appalling. Lights were on everywhere 24 hours a day, the office tower block was lit day and night, and there was a huge heating plant at one end of the factory and an equally enormous air conditioning plant at the other. Both of these ran flat out, apparently in competition with each other.

From our Miami flight, our team was met at the airport on a Sunday by the TRW management who had come in specially to help us get our inspection base set up. By the end of that meeting, chemicals and special dye penetrants, along with the necessary tankage and post-use treatment, were being trucked in from distant parts. I was deeply impressed by the 'can-do' attitude and sheer energy directed at helping us; it was in complete contrast to the stifling bureaucratic approach so typical in the UK at the time. All this organisation was done entirely by phone; paperwork must surely have followed. I was given the exclusive use of an empty ultrasonic immersion inspection tank, with a floor-mounted rotary table, in which I sat with table and chair for my work, and, by the Tuesday, we were able to inspect discs, of which there was a stack waiting for us.

The TRW management were very generous hosts, and we were taken out to a variety of restaurants each day at the end of play where they picked up the tab. Lockheed had assigned an engineer to our team for the whole visit, and he drove us about in a hire car. We were all staying at the Ramada Inn, and I was impressed by the size of the address, let alone the motel itself! On returning from one of the more distant watering holes, I noticed that car lights appeared to be approaching fast on our side of the road and drew the driver's attention to it. Without any hesitation he took our car straight across the grass strip separating the two lanes of the dual carriageway to put us back onto the correct side.

I'm not sure how the mistake was made but I'm so glad I spotted it, having earlier fallen asleep at the table while we were eating!"

DAVID MARSHALL (Former Business Development Director, Aerospace Group)

The RB211 engine and some noise aspects

"In 1971, I was in the Noise Department, working for Mike Smith. We were all based at Hucknall which then had a large open-air test bed for engine noise measurements and the VC10 flying test bed. At around 10am on the morning of February 4th, 1971, all employees at Hucknall were gathered together and we were addressed by Harry Malley, then the site manager. He read out a statement to the effect that the company had become insolvent and was now in the hands of a receiver. He instructed that all employees should continue in their work except that no money was to be spent or committed for the RB211 programme. Further announcements would follow. In the coming days it became clearer that there would be redundancies, but these were made on a last in, first out basis. As I had joined the company in 1961 as a university apprentice, I was in the clear. In fact, I seem to remember that enough voluntary redundancies were offered to reduce very substantially those being forced out, but who wanted to stay.

In relation to the RB211 programme itself, 'noise' became a crunch issue for both the engine and aircraft. The L-1011 was the first aircraft subject to new FAA noise rules from the start. Accordingly, the first noise measurements at Hucknall were closely monitored and extrapolated to predict the aircraft noise against these requirements. The conclusions were alarming, suggesting the aircraft would significantly exceed the limits, preventing its certification. A huge programme started on both sides of the Atlantic to try and solve the problem. However, the most significant break came from the VC10 flying test bed. Flush-mounted microphones in the RB211 intake showed that the intense tone of the fan under static conditions disappeared almost completely by around 60 knots during take-off. It became understood that this was because the intake developed smooth streamline flow from forward speed. Boeing had put 'blow in' doors on the Boeing 747 intake with the Pratt & Whitney JT9D and so did not get this effect. When the first L-1011 flew, those at Palmdale were astonished at its low noise and I believe John Bush worried that there had been an engine shut down!

One of the '1971' effects was to move the Noise Department to Moor Lane with the rest of engineering. I therefore had more regular contact with the seniors and one day had to brief Ernest Eltis. During this, he commented that he wanted to send a liaison engineer to work with Airbus in Toulouse. I volunteered and got the job and went there with my family at the beginning of 1973. By this time the UK had quit the programme at government level, and the RB211 problems, even before 1971, had convinced Airbus they needed another engine supplier, choosing GE and the DC-10 power plant.

Accordingly, Rolls-Royce was not flavour of the month and the only reason for talking to us was a British Airways requirement for a new 200-seat aircraft as a Trident replacement. Airbus believed they would need a Rolls-Royce engine. In the end, of course, BA chose the B757, but we had our 'nose' in the door.

Later, in 1974, I came back to Derby and was asked by Alan Newton, then commercial director, if I wanted to join 'Sales'. However, my remit with Airbus continued and I teamed with Arthur Howes, then in Airbus sales to UK potential customers. We worked together to persuade Rolls-Royce and Airbus to support continued technical work which led to the company agreeing to fund engine pylon design work at BAC. Airbus wanted to showcase this agreement at the Paris airshow in 1975 and it was agreed that it would be publicly signed by the Chairman, Kenneth Keith for Rolls-Royce, and Bernard Lathière for

Airbus. It did not bode well on the day as Lathière was fairly 'mellow' by 3pm and tried to kiss Keith, French style, only to be told that this was still an arrestable offence in the UK. Roger Béteille, then Airbus CEO, then walked into a glass panel thinking it was open space!

But in time all came good. By the time we got the Trent and Airbus to the A330 we had the most competitive combination and set the seal for today's success where we uniquely power the A350, and Airbus is a bigger customer than Boeing. Sometimes you have to be very patient, but the lesson is, never give up!"

IAN TUCKER. MEMORIES OF THE RB211 FLIGHT TEST DEPARTMENT (EARLY 1970'S), FILTON, BRISTOL SITE

"As a result of the financial crisis which befell the company in 1971, part of the overall cost cutting measures involved the cessation of engine flight testing at Rolls-Royce Hucknall site and the transfer of current projects to Rolls-Royce Filton site. This major project change covered the transfer to Bristol of all RB211 flight testing and development in the VC10 flying test bed under the appointment of Bristol Flight Test Engineer, John Dickenson, and supported by Derby. A new team was set up including development, performance, systems and electronic engineers, which I became a part of, to carry out the necessary work to continue support of this prestigious project. Flight testing of the RB211 in the VC10 continued at Bristol from 1972 until 1975 when the aircraft was finally ferried to RAF Kemble for decommissioning. During the period at Bristol, significant milestones were achieved in respect of refining the RB211 engine & performance, handling, relighting, noise and vibration monitoring, safety and mechanical reliability properties in order to meet commercial, contractual and CAA /FAA certification criteria. Flight testing involved many hours of performance monitoring and recordings at various altitudes and airspeeds. Additionally slam and reslam engine accelerations with various engine configurations/altitudes and airspeeds were carried out to establish satisfactory engine response and surge margins at all flight and ground level conditions. Since the ability of any jet engine to satisfactorily relight in flight (carry out an airstart) is an essential feature of meeting safety criteria, much time was allocated to exploring and establishing that the RB211 installed in the VC10 met placarded specifications. I flew as a flight test engineer on 70 of the VC10 RB211 FTB test flights whilst the aircraft was based

at Filton. No doubt significantly more flights occurred when I was not present; Flight testing was not without its perilous moments, and one such incident occurred in August 1972 when an intentionally deactivated and locked out thrust reverser on the test engine decided to spontaneously deploy during the aircraft & initial climb to altitude! The crew experienced a loud bang and the aircraft shuddered and yawed to port. Visual periscope inspection into the front of the engine confirmed that all the cold duct thrust reverser petals had actually deployed and consequently the aircraft was attempting to maintain flight with essentially an eight feet diameter flat plate acting as an airbrake! The RB211 engine was of course immediately shutdown and even with full power selected on the aircraft & two starboard engines, an alarming rate of descent could not be checked due to the heavy all up weight of the fully fuelled aircraft. It was therefore necessary to dump a huge amount of fuel in order to lose a significant amount of weight and thus check the aircraft & rate of descent. Thankfully this was successfully carried out and level flight re-established. Test pilot Dennis Witham had already ordered the crew to check their safety equipment and sea survival rafts in the event of us needing to ditch in the Bristol Channel. Thankfully through both pilot quick thinking and professional flying skill, a disaster was averted, and the aircraft landed safely back at Bristol. The cause of the thrust reverser malfunction was soon identified, and procedures immediately put in place to prevent any reoccurrence. In 1976, I was seconded to Palmdale California to join the team for further engine development leading to FAA/CAA certification of the RB211-524 in the Lockheed L-1011 Tristar. At the same time a colleague also from Flight Test Department Bristol was seconded to Boeing in Seattle to assist in installation trials and certification testing of the RB211-524 in the Boeing 747. At Palmdale, under the local leadership of Rolls-Royce Inc. manager Ron Jackson and his team, much engine development work was carried out alongside Lockheed aircraft testing towards type certification trials in Tristar 1001. Whilst we at Rolls-Royce wrestled with various technical. issues including establishing successful engine handling surge margins at altitude at various VIGV (variable inlet guide vanes) settings, for example, Lockheed continued to progress their designated aircraft tests at the same time, including wet runway, heavy braking and simulated engine failure on takeoff, most of which required the use of extra-long runways. This was carried out generally at Edwards Airforce base which was equipped with a particularly long runway. From engine testing point of view, any particular test we might have scheduled to carry out on a particular flight may have only taken 15- or 20-mins flying time, but to complete all the Lockheed aircraft test items on that same flight, Tristar 1001 (our test aircraft) may have needed to be airborne in excess of 12 hours, and so we often found ourselves airborne for many hours at a time. As increasing numbers of production aircraft were being delivered to customers, occasionally in service engine issues arose, particularly problems with ground starting at various climatic conditions, and so I was dispatched with others to monitor and record actual start up data with Air Canada and TWA scheduled flights. The remainder of my time engaged with RB211 engines in -200 and -500 TriStars involved assisting British Airways with annual Certificate of Aviation renewals, particularly carrying out inflight restarts, an essential feature which needed to be satisfactorily demonstrated."

COMMENTARIES BY FORMER LOCKHEED EMPLOYEES

MIKE KERR (FORMER LOCKHEED REPRESENTATIVE, BASED AT ROLLS-ROYCE, DERBY 1970–1973)

My Rolls-Royce RB211 Experiences

"After growing up on our Iowa farm, I spent four years (1962–1966) at University of Northern Iowa where I majored in Mathematics, after which I received my Draft notice to serve in Vietnam. However, to have an opportunity for Officer Candidate School (OCS), I elected to enlist in the US Army. I graduated from the Fort Sill Oklahoma OCS in June 1967 as a Lieutenant in field artillery and from there I spent a year (July 1967–July 1968) in Fort Bliss, Texas, learning automatic weapons. That was followed by a year in South Vietnam located on the DMZ (demilitarized zone) which divided North and South Vietnam. I left Vietnam in August 1969 at the age of 24.

After three months of vacation, on October 26th, 1969, I was hired by Lockheed Corporation, located in Burbank, California. At that time, Lockheed was under US Defense contracts for the U-2, SR-71, P-3, and other aircraft and aerospace programs critical to the US national defense strategy. It was then starting development of the L-1011 TriStar airbus – a wide-bodied commercial airplane program which was in direct competition with the McDonnell Douglas DC-10. The Rolls-Royce RB211 jet engine (in competition with General Electric and Pratt & Whitney) had been selected for the L-1011. All three engine manufacturers were at various developmental stages.

I joined the Lockheed procurement organisation which was responsible for the Rolls-Royce subcontracted effort and included co-ordination with Rolls-Royce for the supply of Lockheed-supplied accessories which were installed in the RB211. At this early stage, the RB211 and the majority of the L-1011 program were in a 'research and development' stage with a complex fixed price contract established under a development and production program agreement.

On February 4th, 1971, shortly after a joint Lockheed/Rolls-Royce executive review, as a result of significant developmental cost overruns, Rolls-Royce determined it could not sustain associated financial losses and declared bankruptcy. This announcement took Lockheed corporate leaders by surprise and immediately also placed Lockheed Corporation in financial jeopardy itself. At the time, Lockheed and McDonnell Douglas were in fierce competition for both United and American Airlines orders. It came down primarily to who could meet the delivery requirements. There was an immediate sense of shock/panic within the Lockheed Burbank employee base, with many jobs at stake. The Lockheed Procurement Management Team consisted of approximately twelve staff, headed by Ed Brewster, Procurement Director for Rolls-Royce-related Procurement. They included Robert Paul, General Procurement Director, Bud Hale, Manager Rolls-Royce RB211 Subcontract and Larry Fries, Procurement Supervisor, with senior staff Lee Kysar, Mike Wargan, Larry Dewes and 'beginner material associates', Fred Rakeness, Mike Etter, Ed Conn and me, plus a few others. Our contemporaries instantly determined us to be guilty of mismanaging the Rolls-Royce subcontract, or at least that is the way it felt. Up to this time, the working relationship between the two companies was a fairly typical 'arm's length' contractor/subcontractor relationship. Lockheed was pushing hard as L-1011 sales were in the balance and indications were that RB211 research and development efforts were not proceeding to plan and/or were not being accurately conveyed. L-1011 manufacturing was also being faced by numerous other challenges for this first-of-a-kind airplane. The announcement that day resulted in a challenging, sad and demoralised time, but it was also clear that extraordinary measures must be taken to save the program. So, we took a breath and went to work double-time, so to speak.

The Rolls-Royce announcement was followed five days later (February 9th, 1971) by a severe 6.7 earthquake at 6am, centered in Sylmar, California, 14 miles from the Lockheed Burbank location. (This earthquake is better known as the San Fernando earthquake.)

The L-1011 assembly line was located in Palmdale, 50 miles east-north-east of Burbank. There were numerous warehouse buildings on the Burbank property containing various aircraft parts and most tumbled from the multi-storey storage racks, making a terrific clatter of falling metal and damaging parts. Fortunately, no one on the Lockheed Burbank site was hurt as the morning shift was changing and few were occupying the buildings. I was starting to get up to get ready for work but was literally thrown out of my bed onto the floor. My apartment was located 14 miles from the epicentre. I got up from the still-shaking floor and pulled open the sliding glass door which was located on my second-floor balcony overlooking the swimming pool. I raced outside, momentarily stood there, and then realised the shaking building might cause the balcony above me to collapse. As I was contemplating that, the pool created its own tsunami. Water went from pool level to a wave over 12-feet high which totally drenched me, head to toe as I stood in my underwear. Two feet of water instantly left the confines of the swimming pool. It was a rude awakening. I ran quickly to the hallway as it seemed the building would collapse. Electrical wires leading from the building to transformers were snapping. All the occupants were racing out in their nightwear to get away from the building. We headed for North Hollywood Park across the street. Trees were being uprooted and falling. There was concern that a nearby dam would break and flood the area. It didn't happen, however, numerous buildings collapsed, homes, hospitals and businesses were damaged, freeways and overpasses collapsed, and 64 people lost their lives.

That event, shortly following the Rolls-Royce announcement, added another level of turmoil. The Lockheed and Rolls-Royce executives and US/UK governments raced to try and rescue the situation. One form of communication with Rolls-Royce was via a telex machine and telephone (at that time international telephone calls ran a little over $3 per minute, with a delay in voice transmission). Time was money. We used a hand calculator to run the numbers. We would either dictate or write the communication text in longhand. The written text would then be given to our typing pool group. Once the communique was typed, we would proofread it and then it would be sent.

After considerable effort, Rolls-Royce was restructured, and the UK government managed the receivership. In parallel, Lockheed and Rolls-Royce personnel worked around the clock to restructure the RB211 contract and its pricing. By a single vote, Lockheed received the first ever US government federal loan guarantee ($250 M). It was *also clear to each party that we needed to work as a team, be open and transparent with each other and to our respective government overseers. This relationship started at executive level and continued down to all levels, engineering, manufacturing, quality control, scheduling, contracts and finance.*

When Rolls-Royce made its announcement, it became obvious to all parties that both companies would need some form of government intervention to maintain company solvency, very unusual in a commercial environment.

Lockheed immediately expanded its Rolls-Royce office staff from five people to 28. The expanded office included a Procurement Director (C.H. Roha), engineering technical experts, quality control personnel, planners and schedulers, and procurement subcontract personnel (me and two others) to assist the Burbank procurement organisation to monitor status and negotiate pricing and specification changes, and to schedule revisions to contracts with Rolls-Royce, Lockheed suppliers located in England, Scotland, France and Switzerland, and US suppliers that provided RB211 accessories.

Each day, the Lockheed office director would report the detailed engine status, technical and schedule issues, and personnel needs to the then, Burbank Lockheed President and manufacturing expert (Archie Folden). Lockheed corporate executives also became frequent visitors: Dan Haughton, Chairman; Roy Anderson, Chief Financial Officer and later Board Chairman; Willis Hawkins and Chuck Wagner, successor Vice Presidents; Kelly Johnson, Lockheed Skunk Works Executive; Tony Lavier, Chief Flight Test Pilot, to mention a few big names in our industry. I was a non-political farm kid surrounded by senior Lockheed corporate figures, all nationally well-known and politically tied to airplane/aerospace endeavors. The livelihood and success for both companies hinged upon open communication and accurate assessments of the RB211 development and production completion. Lockheed pushed its L-1011 production line to the point that simulated jet engines (cement) had to be mounted to wings pending receipt of RB211 engines. Immense pressure was placed on both companies as L-1011 sales were being impacted and the governments of the UK and US were under considerable political pressure. The US government loan guarantee was the first ever for a US commercial company. Unbeknown to me at the time, Lockheed had several critical defense programs which were a key part of the US national security strategy. Lockheed solvency was extremely important!

Then, at age 26, during April 1971, I was thrilled to be asked to travel to Rolls-Royce as a representative to basically chase parts for two-plus years! The US government loan guarantee had not yet passed, but I was willing to swim back or pay my way back if it didn't. I was still a bachelor, so flexible with respect to relocation and hours on the job. I had served as an Army Lieutenant in Vietnam, travelled to Australia, Japan, and to 25 states in the US, so travel was a major perk incentive at my young age. I was very excited with the prospect of living in England, where my dad had also served as a World War II Army Airforce pilot. After memorable going away parties, I left for England in July 1971. I had no idea how to get to Derby but was not intimidated in the least–just excited!

Lockheed had given me $1,000 for travelling expenses to get to Derby. I was still packing my suitcase as a friend drove me to the Los Angeles LAX. In fact, what I couldn't pack, I tossed out of the window on the Interstate 5 freeway to the airport. (Packing was a last minute 'priority' for me because of going to too many 'leaving parties!) I left my 1969 Dodge Charger with my brother as I didn't realise it could have been shipped to England. Mistake No. 1 as it was (and still is) worth quite a fair amount, but the Charger steering wheel was on the wrong side for UK driving.

I had to catch a train to get to Derby. Up to that time, I had never had a ride on a train, but the transportation system in England was great, so I found my way. One of the office representatives, Mike Etter, gave me a ride from the Derby train station to the Pennine Hotel. This is where we also rented cars until we purchased our own personal vehicle. As it turned out, Lockheed eventually purchased five rental cars as they were totalled on the way out of the rental location–usually a head-on collision within the first 50 feet of the exit. Driving on the left side of the street was truly a learning experience for us Yanks! Another adjustment was the weather. Several families stayed only a few months as the transition from nearly 80% California sunshine to considerably fewer sunshine days in the UK was too difficult. But the weather didn't bother me as I spent many hours in the Rolls-Royce facilities or at the most excellent pubs in the Derby area, especially the John Thompson Inn and the Crewe & Harpur. I was in the prime of my work/party life and made the most of it, but the mission, my role to support Lockheed and Rolls-Royce to the best of my ability was foremost. It was all about the mission. 'Get the job done' was instilled into me on the Iowa farm and in the Army. It stuck with me.

The Lockheed executives treated me as a son. I didn't recognise it at the time, but it was an honour, I saw it as simply doing what I was paid to do. There were a good number of around-the-clock days. In addition to daily scheduling and the tracking of Lockheed supplied accessories for the RB211 suppliers with my Burbank counterparts or directly with supplier representatives, e.g., Sundstrand, integrated drive generator; Walter Kidde, fire detection systems; Hamilton Standard, high pressure shut-off valve; to name a few, I also helped to clarify the status of certain internal Rolls-Royce component deliveries.

Also, when RB211 engines were completed and accepted by the Lockheed quality representative (John Grimm), I would notify the Lockheed Burbank Procurement Director (Ed Brewster) when we needed a Saturn C-130 aircraft at the East Midlands Airport to load three engines. Time was of the essence as airplanes were awaiting engines and L-1011 deliveries were critical to sales. When the airplane and engines arrived, no matter the time of day, and often after midnight, I had to witness the loading as ownership and associated insurance transitioned from Rolls-Royce to Lockheed as the RB211 moved from the Rolls-Royce transportation truck to a portable loading dock situated between the truck and the C130.

When the transition was completed, I signed a 'transfer off' form on behalf of Lockheed Corp. The one exception occurred when a cable from the wrench inside the C-130 airplane used to pull the engine from the Rolls-Royce truck to the interior of the airplane snapped and damaged a RB211 engine cowling. The cowling was later replaced at Lockheed Palmdale facilities at a cost of, as I recall, $160,000.

During my first year and a half at Lockheed Burbank, I had three union grievances filed against me. I just didn't understand unions and I just did what was needed to get a job done, but I soon discovered there were rules! Neither Lockheed nor Rolls-Royce could afford a shutdown because of a union violation. Then I learned that there were different unions within the Rolls-Royce RB211 manufacturing facility. From my Lockheed grievance experience, I knew it wasn't wise to argue with a union steward or lift and move something I shouldn't. My level of patience was not great as I was mission centric. However, I started to learn that a level of discretion and patience was a good thing. It served to dispense a little political acumen plus it alleviated some aggravation as it seemed to me that my English counterparts spoke in a different language! That was another adjustment – mostly fun, but sometimes just a little difficult to bridge my understanding of US English to the English translation!

For example, about a year into my residence, after a long day I stopped at a local pub and had but one fine ale, then left. There was construction on the local road and for a short distance over a bridge, so traffic was limited to one lane, my lane. As I was proceeding over the bridge, a mini came at me and a head-on collision ensued. I was driving my fine Ford Capri and was on the proper side of the road. The police questioned both of us. The two gents in the mini had also been to a pub for a bit too long and were, well, somewhat incoherent. But I had to give my version of the accident to the Police, and I explained that, "I was on the right side of the road". That added to the confusion as my Yankee accent was a giveaway that I wasn't from these parts and obviously should not have been on the right side of the road.

After a fair bit of explaining, I finally translated to English and then explained that I was on the proper side of the road, e.g., the left side, which was not the right side?"

Sadly, Mike passed away a year after sending me his recollections. We had some very interesting conversations, and he proved a considerable help and assistance to me during my research period. I am grateful and privileged to have known him, albeit too briefly.

JAMES FITZGERALD (Lockheed Engineering Flight Test Department)
Lockheed TriStar 3 -Flight Test Aircraft – Trip to Paris, 1971

"In 1971, things were looking bleak for the L-1011 program. Our engine manufacturer, Rolls-Royce, had declared bankruptcy, and our airplanes were slow in coming off the assembly line due to a steep learning curve in manufacturing. We were in quite a race with our competition to sign up new customers. A decision was made at corporate headquarters to send a TriStar to the 1971 Paris Air Show.

The goal was to send the number four TriStar, which would have been the first one with a fully configured passenger interior. With the Rolls-Royce problem, this became impossible. We had three airplanes flying in the flight test program, and the company committed to send TriStar 3 to Paris. The airplane had a partial passenger configuration, but also quite a flight test instrumentation system installed. Most importantly, it contained three functional RB211 engines, of the nine which we had at the time, the other six being installed on flight test airplanes one and two.

I'm not sure on what date TriStar 3 made her first flight, but my first flight on the airplane was flight number six, on May 25th, 1971. Three days later, on flight number nine, we left Palmdale and headed for the Paris Air Show. Stops at Minneapolis and Montreal led to our landing at Goose Bay, Labrador, on May 30th. After servicing and refueling at Goose Bay, we took off to head across the North Atlantic. Tom Laughlin was our navigator, and I had responsibility to keep up with data from our lone inertial navigation system aboard, to complement Tom's efforts.

Soon after lift-off, a 'boom' sounded on the left side of the airplane, and John Radich reported he saw a momentary flame from the exhaust. Looking out of the windows, it was obvious that a small access panel had departed the number one engine, (See Image 42). The crew shut down the engine, and I can't remember if we dumped fuel, but suspect that we did. Here we are at a higher gross landing weight than ever before in the test programme, and that limit had not even been determined at this point. We were also a two-engine TriStar. We returned to Goose Bay and landed successfully.

Once back on the ramp, the cowlings were opened on the engine, and large chunks of thick metal fell out onto the ground. A gaping hole was observed in the RB211 gear case and oil was dripping on the ramp. It started to rain about then and kept on for a long time. We had Rolls-Royce personnel with us, and they quickly determined what must be done to correct this problem. We needed a replacement gear case.

Simple, right? There were only nine of these in existence in the USA. at the time, all on our flight test airplanes. This is when the real character of our company (back then) really came into play. No one, at this point in the history of the RB211, had ever removed and/or replaced one of these gear cases, which not only provided oil lubrication to the engine, but ran all the auxiliary equipment such as hydraulic pumps and electrical generators. They weighed about 800 pounds. Many quick phone calls were made and immediately work began in Palmdale to remove the gear case from one of the other flight test airplane engines. At the same time, the same work began on the ramp at Goose Bay. Kelly Johnson's JetStar was reserved to fly the replacement gear case to Goose.

I had forgotten that a 'Plan B' had been initiated by the company. An engine was removed from one of the other flight test airplanes in Palmdale, and Saturn Airways was contracted to ferry that engine to Goose Bay in the event that an engine change was necessary. I found a photo of that Saturn C-130 on the ramp at Goose, (see Image 41).

Amazingly, this all happened within 24 hours. When the JetStar arrived at Goose, the replacement gearbox was unloaded, and installation started immediately on TriStar 3's engine. Rain continued and it taxed the improvised tent erected around the work area. Once the installation was complete, the real fun started, and this is what I most remember from this adventure.

We started engines and taxied to an area where we could run the number one engine up to power, which we did. We taxied back to our parking area, opened the cowlings, and the Rolls-Royce guys pulled the oil filters. They were about half full of metal chips. Changed filters and engine oil, repeated the taxi/run up and checked filters. Now only about a quarter full of chips. Again, we changed filters and oil and repeated the process. This time there were only a few chips and the Rolls-Royce folks pronounced that we were once again airworthy.

On May 31st we took off again from Goose Bay, headed across the North Atlantic for Paris, by way of Shannon, Ireland. We crossed the Atlantic at night, and the stop in Shannon is deserving of another long dissertation. We made it to Paris on June 1st in time for the show. Hopefully we signed up some new customers."

JOHN PLOWMAN (Former Lockheed Employee)

"After graduating from the City University in 1970, I joined Court Line in 1971. After Court Line folded in 1974, I joined Cathay Pacific in Hong Kong. In 1979, I joined Lockheed, firstly in Field Service in London and later in Product Support in Burbank, California. I retired from Lockheed Martin in Marietta, Georgia, in 2006. I should also mention I worked for Rolls-Royce for about six weeks in the summer of 1966 as a student apprentice at Derby and I have always felt this was a major plus on my C.V.

Before joining Court Line, I was aware that the company was planning on operating the L-1011, but at the time, this seemed a far-off dream. Not long after joining, all maintenance staff were called to a meeting to be told the acquisition of L-1011s would be significantly delayed. This did not unduly bother me as the airline was operating 13 BAC1-11s and appeared to be doing well. The fact that Lockheed itself was having serious financial problems was only briefly mentioned and it wasn't until later, in August 1971, that it was able to resume production of the L-1011. My guess is the receivership of Rolls-Royce, and the potential insolvency of Lockheed delayed the delivery of L-1011s to Court Line by a year. Court Line received the eighteenth and twenty-fourth production aircraft. At the time of Court's demise, Cathay Pacific had ordered L-1011s and later received the one-hundred-and-eighteenth and one-hundred-and-twenty-second deliveries.

Where I feel the Rolls-Royce receivership directly affected customers was in supporting RB211 deliveries and funding the resolution of technical and mechanical issues that began to be manifest as the engine entered airline service. I suspect that after receivership, funds and budgets were tight and the development of costly fixes was delayed or drawn out, at times over years. For example, serious mechanical issues with the engine thrust reverser system (specifically the hot stream spoiler) were not entirely resolved until about 1976 and by that time it had been officially decided to permanently delete that part of the thrust reverser system. Other issues that seemed, from the point of view of a flight-line mechanic/inspector/supervisor, to take an inordinate time to fix were problems with the engine airflow control system that resulted in compressor surges and sometimes damage to the IP compressor VIGV system, turbine blade failures, and general unreliability of the fuel flow regulator (FFR) made by Lucas. I should mention that engine problems on the L-1011 were not the only problems on the aircraft in the early years. Main landing gear tyre failures and the shattering of the outer layer of windshield glass were costly and highly disruptive problems at the time.

By the time the RB211 was in service with Cathay Pacific in late 1975, most of these engine issues were slowly being resolved. Engine life in terms of flying hours was dismally low by today's standards. I remember a Kai Tak flying club party in about 1977 to celebrate an engine achieving 2,000 flying hours and I also remember TWA and Delta achieving 20,000 hours in the mid-90s."

25. Sir Denning Pearson, Rolls-Royce Chairman, 1969.

26. Sir David Huddie, Rolls-Royce Managing Director, 1969.

27. Dan Haughton's first visit to Derby after signing of the RB211 contract for the Lockheed L-1011 aircraft. Sir David Huddie (left), Rolls-Royce, with Dan Haughton, Lockheed Chairman (centre) and Charlie Hunt, Rolls-Royce Works Convenor (right).

28. Early design variant, RB211-06 engine.

29. An early RB211-22B engine with Hyfil fan blades.

30. The first RB211 engine on test at the Sinfin A site, 1968.

31. RB211 engine in a test cell.

32. RB211-22B on the No. 10 test bed at Rolls-Royce Hucknall.

33. Rolls-Royce Hucknall No. 9 test bed 1970, RB211 ground crosswind test.

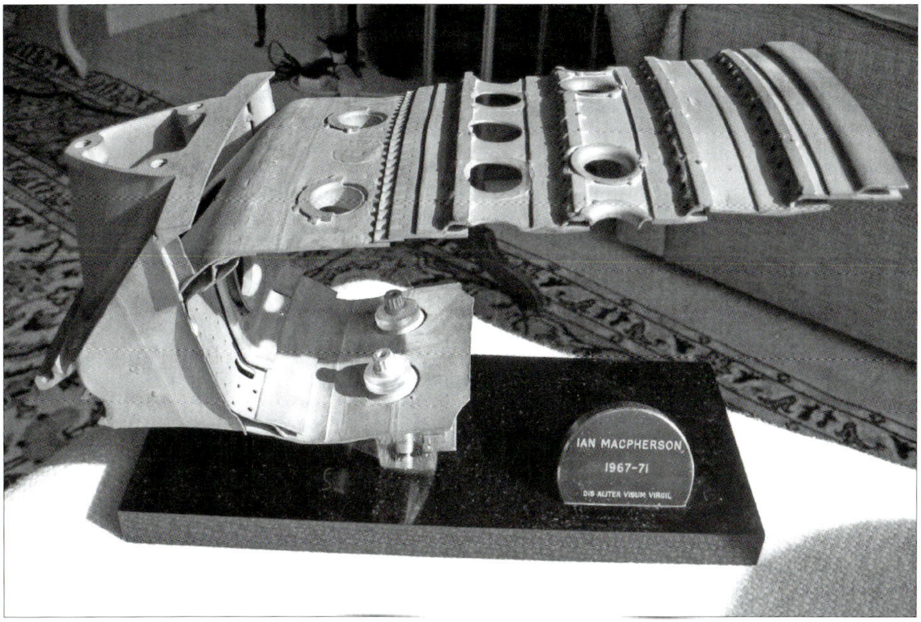

34. RB211 development part.

35. Eastern Airlines Lockheed TriStar, fly-past of the Rolls-Royce Hucknall site.

36. RB211 Civil Aviation Authority engine certification ceremony, 1972.

37. Celebrating the 555th production RB211 engine for Lockheed.

38. Rolls-Royce engine build and assembly shop.

COST COMPETITIVENESS

Mr J. A. Rigg Financial Controller Aero Engine Division

TO enable us to maintain, let alone increase, our share of the future world market for aero-engines it is not sufficient merely to be technically superior to our competitors. We must also offer a real cost advantage to the airlines if they operate our engines. Both these aspects of our competitiveness are interlinked, since it is no use designing and manufacturing a high-quality product which no-one will buy.

One of the most significant elements of cost in any product is the cost of not doing the job 'right first time'. I am not just thinking of the obvious things like scrap and rectification costs in the manufacturing areas, but also those other error-correcting activities which are much less easy to put a value on — amending drawings, re-typing letters, re-running computer programmes, revising schedules and so on. The list is endless, and there is not a single person who could not contribute in some way to reducing these unnecessary costs. In so doing we would all be helping to improve the quality of the Division's overall performance, and increase our prospects of competing successfully for the business which will keep us occupied for the next 15 years.

Mr J. A. Rigg

39. Jim Rigg, Rolls-Royce Finance Controller; RB211 cost control campaign 1968.

40. Inspection of turbine blades, circa 1970, precision cast facility.

41. RB211 engine cowl open, emergency landing, Goose Bay, May 1971.

42. TriStar 3 RB211 engine – missing access panel, May 1971.

43. Rolls-Royce workers at the Castle Donington site (a happy band of men).

44. 'Beam me up Scotty': Rolls-Royce workers at Rolls Royce, Castle Donington site

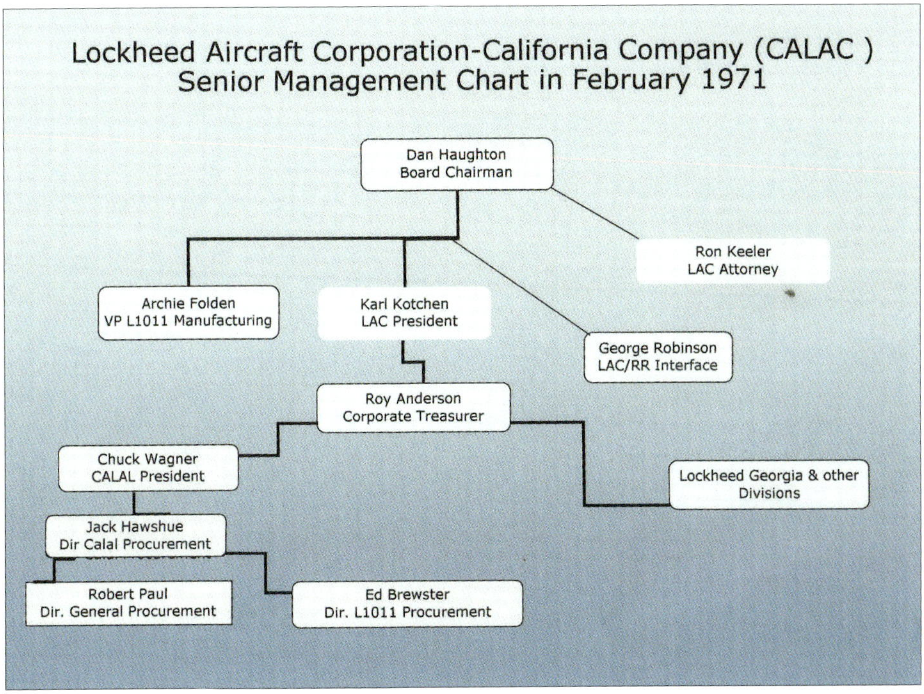

45. Lockheed Aircraft Corporation, board and senior management chart, February 1971.

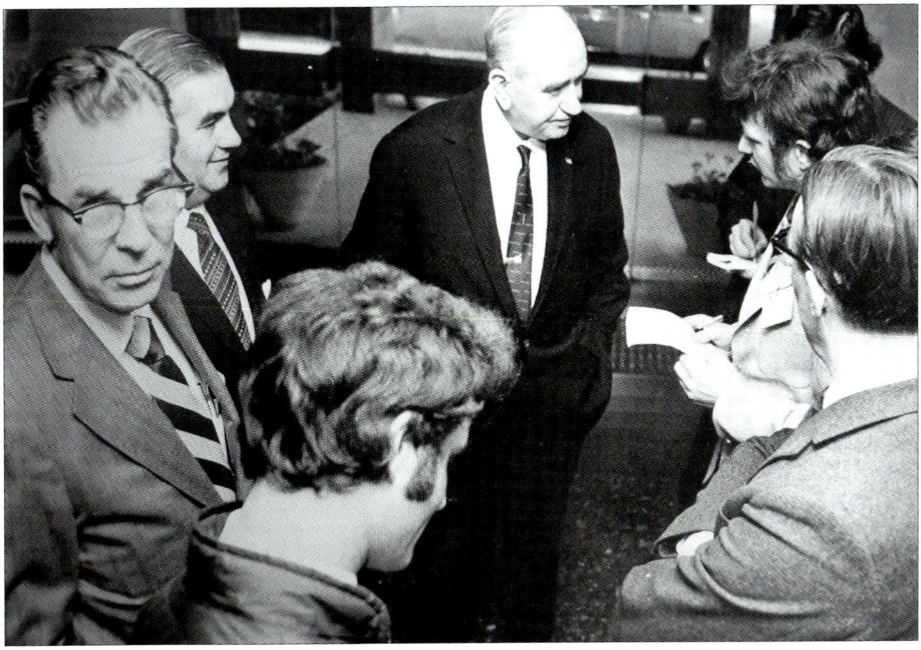

46. Lockheed Chairman Dan Haughton (centre) in discussion with Rolls-Royce employees.

47. 'Little and Large', comparison of aircraft sizes: Lockheed single engine Vega, (wooden) built in 1929, against Lockheed L-1011 TriStar, built in 1971.

48. Lockheed TriStar production lines.

LA293 02 09221

7 RJ

C.H. ROHA

04/05/72

LOCKHEED DELIVERS FIRST L-1011
TO EASTERN AIRLINES AT PALMDALE

PALMDALE, CALIF., APRIL 5-- LOCKHEED DELIVERED THE FIRST OF ITS NEW ROLLS-ROYCE POWERED L-1011 JETLINERS TO EASTERN AIRLINES TODAY AND AN AIRLINE CREW FLEW IT AWAY TO MIAMI FOR A MONTH OF PILOT TRAINING BEFORE IT GOES INTO AIRLINE SERVICE.

IT WAS A DRAMATIC MOMENT FOR HUNDREDS OF LOCKHEED EMPLOYEES WHEN THE SLEEK, BLUE AND WHITE LUXURY JET TAXIIED AWAY FROM THE COMPANYS ASSEMBLY AND FLIGHT TEST CENTER HERE.

LOCKHEED BOARD CHAIRMAN DANIEL J. HAUGHTON, WHO IS IN ENGLAND TODAY, VISITED THE PALMDALE FLIGHT LINE MONDAY TO WISH BON VOYAGE TO THE EASTERN CREW PRIOR TO THEIR DEPARTURE IN THE L-1011.

"THE DELIVERY OF THE FIRST PASSENGER-CARRYING L-1011 TO EASTERN AIRLINES IS ANOTHER GREAT MILESTONE IN THE TRISTAR PROGRAM," HAUGHTON SAID. "TWENTY-ONE ADDITIONAL TRISTARS WILL BE DELIVERED TO OUR AIRLINE CUSTOMERS THIS YEAR.

"I REMAIN CONFIDENT THAT THIS AIRPLANE IS GOING TO BE ITS OWN BEST SALESMAN AND THAT EVENTUALLY LOCKHEED WILL OBTAIN ITS SHARE OF THE TRIJET MARKET WITH BOTH TRANSCONTINENTAL RANGE AND EXTENDED RANGE TRISTARS," HAUGHTON SAID.

"I ALSO AM CONFIDENT THAT THE LOCKHEED L-1011 IS THE BEST OF THE WIDE-BODIED JETLINERS AND THAT IT WILL BE SERVING THE AIRLINES AND THE PASSENGERS WHO WILL FLY IT FOR MANY YEARS TO COME," HE ADDED.

EASTERNS FIRST L-1011 WILL BE USED INITIALLY FOR FLIGHT AND GROUND CREW TRAINING, WITH THE START OF SCHEDULED PASSENGER SERVICE EXPECTED IN ABOUT ONE MONTH.

LOCKHEED EXPECTS TO RECEIVE THE FEDERAL AVIATION ADMINISTRATION S CERTIFICATE OF AIRWORTHINESS FOR THE TRISTAR IN MID-APRIL. TRANS WORLD AIRLINES ALSO WILL RECEIVE A TRISTAR FOR FLIGHT CREW TRAINING PRIOR TO TYPE CERTIFICATION.

FORMAL TRANSFER OF THE FIRST TRISTAR FROM LOCKHEED TO EASTERN WAS CONCLUDED TODAY WHEN WILLIAM M. MORAN, EXECUTIVE VICE PRESIDENT OF LOCKHEED-CALIFORNIA COMPANY HANDED THE AIRCRAFT S KEYS TO PAUL M. JOHNSTONE, VICE PRESIDENT OF ENGINEERING FOR THE AIRLINE.

AN EASTERN DELIVERY CREW BOARDED THE AIRCRAFT SHORTLY AFTERWARDS AND FLEW IT TO THE AIRLNES MAIN BASE AT MIAMI. THE CREW INCLUDED CAPTAINS PAUL SLAYDEN, THAD ROYALL AND GASPAR GARCIA AND SECOND OFFICER HAROLD SNELL.

FIRST DELIVERY OF THE TRISTAR WAS THE CULMINATION OF A DEVELOPMENT PROGRWM THAT BEGAN FOUR YEARS AGO. THE L-1011 IS A THIRD-GENERATION JETLINER, INCORPORATING ADVANCES IN TECHNOLOGY AND A PASSENGER CAPACITY MORE THAN DOUBLE THAT OF THE FIRST GENERATION OF COMMERCIAL JETS. MAJOR BENEFITS OF THE NEW DESIGN ARE A SUBSTANTIAL REDUCTION IN NOISE AND IMPROVEMENTS IN PASSENGER COMFORT, OPERATING EFFICIENCY AND

49. Lockheed telex to C.H. Roha announcing the first TriStar delivery to Eastern Airlines, April 5th, 1972, in readiness for press release.

SEAT-MILE OPERATING COSTS.
FLIGHT TESTING OF THE TRISTAR BEGAN NOVEMBER 16, 1970 WITH MAIDEN FLIGHT OF THE FIRST L-1011. SINCE THEN FIVE TEST AIRCRAFT HAVE FLOWN A TOTAL OF MORE THAN 1800 HOURS.
EXTENSIVE NOISE MEASUREMENTS DURING THE TEST PERIOD HAVE SHOWN THE L-1011 TO BE THE QUIETEST JETLINER NOW FLYING OR SCHEDULED TO ENTER COMMERCIAL SERVICE.
OTHER OUTSTANDING CHARACTERISTICS OF THE L-1011 ARE ITS EXCEPTIONAL STABILITY AND RESPONSIVE HANDLING QUALITIES BROUGHT ABOUT IN PART BY AN INNOVATION CALLED THE QUOTE ALL-FLYING HORIZONTAL STABILIZER. UNQUOTE THIS DESIGN CMA NEW TO U.S.-BUILT TRANSPORTS GIVES THE PILOT DIRECT CONTROL OVER THE ENTIRE HORIZONTAL TAIL RATHER THAN JUST THE ELEVATOR.
THE TRISTAR S AUTOMATIC LANDING SYSTEM ALSO HAS DRAWN WIDESPREAD RECOGNITION FROM THE AIRCRAFT INDUSTRY AND COMMERCIAL AVIATION.
MORE THAN 300 FULLY AUTOMATIC LANDINGS WILL HAVE BEEN PERFORMED BY TRISTARS BY THE TIME FAA CERTIFICATION IS ACCOMPLISHED.
SOME TESTING WILL CONTINUE AFTER CERTIFICATION. AMONG THESE TESTS IS THE SIMULATION OF 36000 AIRLINE FLIGHTS ON A STRUCTURALLY COMPLETE TRISTAR ENCASED IN A STEEL FRAME WITH SOME 300 HYDRAULIC JACKS ATTACHED TO ITS EXTERNAL SURFACES.
CALLED FATIGUE TESTING THIS PROGRAM WILL BE COMPLETED LATER THIS YEAR CMA PROVIDING A DEMONSTRATION OF THE L-1011 S RESISTANCE TO STRUCTURAL FATIGUE OVER A 20-YEAR LIFETIME OF AIRLINE SERVICE. THE FATIGUE TEST AIRCRAFT TO DATE HAS MADE MORE THAN 15000 SIMULATED FLIGHTS.

VERN MAC PHERSON
PUBLICITY MANAGER
LOCKHEED BURBANK
213 847 5730 L-1011 5 L-1011 L-1011 L-1011 L-1011 L-1011 16 1970 L-1011
1800 L-1011

L-1011 300 36000 300 L-1011 20 15000

1124A PG

RJ829

#

49a. Lockheed telex, continued.

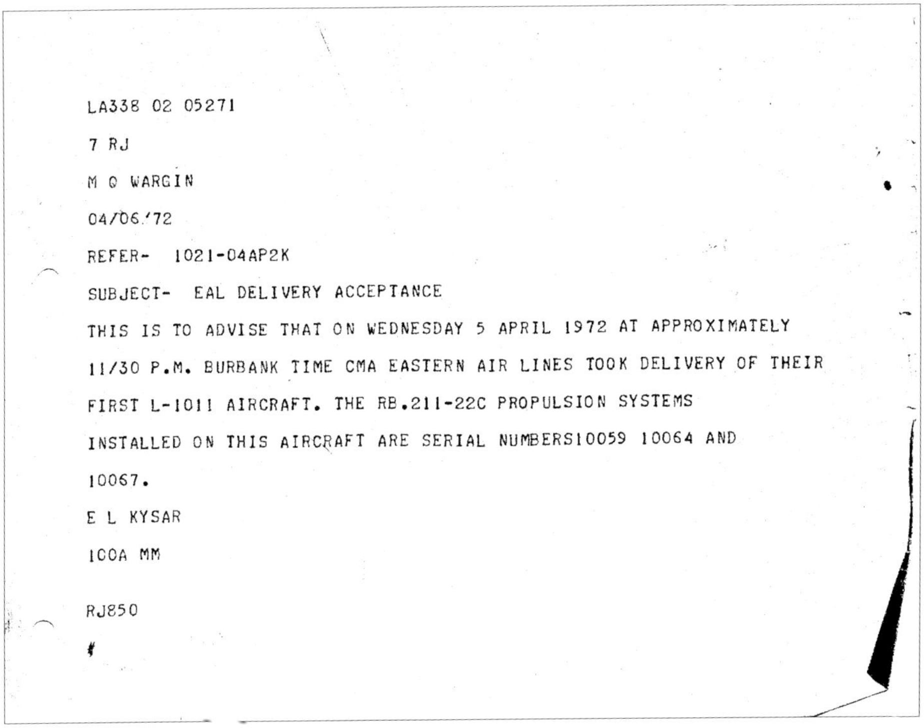

LA338 02 05271

7 RJ

M Q WARGIN

04/06/72

REFER- 1021-04AP2K

SUBJECT- EAL DELIVERY ACCEPTANCE

THIS IS TO ADVISE THAT ON WEDNESDAY 5 APRIL 1972 AT APPROXIMATELY 11/30 P.M. BURBANK TIME CMA EASTERN AIR LINES TOOK DELIVERY OF THEIR FIRST L-1011 AIRCRAFT. THE RB.211-22C PROPULSION SYSTEMS INSTALLED ON THIS AIRCRAFT ARE SERIAL NUMBERS10059 10064 AND 10067.

E L KYSAR

1COA MM

RJ850

50. Lockheed telex dated April 6th, 1972, to M.Q. Wargin confirming the first L-1011 TriStar delivery to Eastern Airlines and advising of delivery of the L-1011 aircraft; RB211 propulsion systems, serial numbers 10059/10064/10067.

PROVISIONAL

19 January 1973

AGENDA FOR MEETING WITH D J HAUGHTON
MONDAY 22 JANUARY & TUESDAY 23 JANUARY 1973

Location: Boardroom, G Block, Moor Lane

8.30 am	1	Fan Disc Failure	
	1.1	Statement of service position	A Rhodes
	1.2	Investigation programme	J F Bush
	1.3	Design action	J F Coplin
	1.4	Visiting team report	W M Hawkins
9.45 am	2	RB 211-22B Programme	
	2.1	Certification programme and performance situation	E M Eltis
	2.2	Status of production of new engines and conversion kits	J O Keir
10.15 am	3	Service Status	
	3.1	Review of the current operating experience highlighting the major problems	A Rhodes
	3.2	Airline support	A Rhodes
11.00 am	4	Engine Life Improvement Programme	
	4.1	Short term improvements	J F Coplin
	4.2	Long term improvements	J F Coplin
11.30 am	5	Tour of Facilities	
		Including inspection of RB 211 certification hardware.	F T Salt
12.30 pm		Lunch	

/continued.......

51. Provisional agenda for the visit of Lockheed Chairman Dan Haughton to Rolls-Royce, January 22nd and 23rd, 1973.

\- 2 -

1.45 pm	6	<u>Specific Air Range Situation</u>	
	6.1	Statement on current position	E M Eltis
	6.2	Future Rolls-Royce and LAC action including NGTE and flight testing	E M Eltis
2.30 pm		Executive Group only.	
		Attendees to be nominated by G Fawn and R Fuhrman.	
	7	<u>Engine Pricing</u>	M Day
		General Discussion	

<u>TUESDAY 23 JANUARY</u>

8.30 am	1	<u>Extended Range L1011-2</u>	
	1.1	L1011-2 launch programme	R Fuhrman
	1.2	RB 211-524 programme	K J Spooner
	1.3	RB 211-524 contract status	
	2	<u>Compatibility of Engine/Airframe Schedules</u>	

<u>A G COLLINS</u>

51a. Provisional agenda, continued, page 2.

DB056 02 05270

7 RJ

04/26/72

M S KERR

REPEAT/ C H ROHA

LOCKHEED/DERBY

EAL MADE FIRST L-1011 PASSENGER REVENUE FLIGHT THIS MORNING CMA APRIL 26 CMA MIAMI TO ATLANTA FOLLOWED BY SECOND FLIGHT ATLANTA TO NEW YORK. THREE MORE SCHEDULED FOR TODAY COLON
NEW YORK TO MIAMI CMA MIAMI TO NEW YORK CMA AND NEW YORK TO MIAMI.
ABOVE INFORMATION AS OF 1.00 PM EST AND ALL REPORTED GOING WELL.

E L BREWSTER

1152A AMC

RJ228

52. Lockheed telex April 26th, 1972, to Mike Kerr, Lockheed regarding Eastern Airlines, confirming the first L-1011 passenger revenue flight.

WASHINGTON OFFICE:
2252 RAYBURN HOUSE OFFICE BUILDING
202-225-5811
ROBERT C. RUBEN
ADMINISTRATIVE ASSISTANT

HOUSE WAYS AND MEANS COMMITTEE
SELECT COMMITTEE ON SMALL BUSINESS

JAMES C. CORMAN
21ST DISTRICT, CALIFORNIA

Congress of the United States
House of Representatives
Washington, D.C. 20515

DISTRICT OFFICE:
14450 ERWIN STREET
VAN NUYS, CALIFORNIA 91408
213-787-1776
IRENE SLATER
FIELD DEPUTY
SARAH ETHEREDGE
FIELD REPRESENTATIVE
NELDA BARRETT
CASE WORKER

September 3, 1975

Mr. M.S. Kerr
Senior Buyer
Lockheed Aircraft Corporation
P.O. Box 551
Burbank, California 91520

Dear Mr. Kerr

The eighth annual Business Opportunities in Federal Procurement and Foreign Trade Conference, held August 28, was very successful, with over 600 participants in attendance.

Lockheed Aircraft Corporation contributed greatly to its success with valuable counseling and specific information on procurement.

Thank you for attending the conference, and I look forward to your participation next year.

Warm regards.

Sincerely

James C Corman

JAMES C. CORMAN
Member of Congress

JCC:is

THIS STATIONERY PRINTED ON PAPER MADE WITH RECYCLED FIBERS

53. Letter to Mike Kerr, Lockheed, from the US Congress, September 3rd, 1975.

COPYRIGHT ACKNOWLEDGEMENTS

Images 25–26, 29–40 all copyright Rolls-Royce Limited, with kind permission of Rolls-Royce Heritage Trust.

Image 27 copyright Rolls-Royce Limited, Courtesy of the Derby Evening Telegraph archive and copy also supplied by Charlie Hunt's family.

Image 28 Courtesy of the late Dr. Ian Macpherson.

Images 41–42 Courtesy of James Fitzgerald.

Images 43–44 Courtesy of Richard Smith.

Image 45 Author from information provided by Mike Kerr.

Images 46, 51, 51a copyright Rolls-Royce Limited, with kind permission of the Rolls-Royce Heritage Trust.

Images 47–50, 52–53 copyright Lockheed Martin, courtesy of the late Mike Kerr.

CHAPTER 5

THE GATHERING STORM – A DETERIORATING FINANCIAL OUTLOOK

COMPANY FINANCIAL CONTROLS

Accountancy tools and change in accountancy policy:

The Five-Year Forecast: The long-term planning document.

These were in their infancy, having been introduced in 1966 for the 1967-1971 period. The forecasts provided information which included escalation factors to allow for the effects of inflation, profit and loss accounts and balance sheets for each of the five years. Five-year forecasts were prepared differently by each division or subsidiary. However, there were no set guidelines as to how these should be presented or what information should be included. The forecast included a dual forecast for the RB207 and RB211 programme. Each forecast would include profit improvement plans such as efficiency savings targets by the DED engineering group. The divisional forecasts were then collated at headquarters by a small team who would add a round sum contingency margin into the consolidation for the group forecast. These forecasts were originally intended to be submitted in June of each year independently of any profit plan. This changed in October 1969 to enable the profit plan of 1970 to be included as the first year of the 1970-1974 forecast.

Five-year forecasts and profit plans at divisional and group levels were subject to scrutiny and approval of the chief executive, main board finance committee and the main board.

Due to the rapidly escalating costs of the RB211 these were then revised more regularly than once per year, to reflect the changing circumstances. However, it was not until the final stages that it became clear to the main board of the serious cash situation the company was in and the impending financial disaster.

The Annual Profit Plan: A short term planning document: Prepared at company headquarters, this involved the development of an interlocking set of budgets that rolled up into a master budget and were revised twice a year in May and

October as to the forecasted results (these two regular revision periods would not change until mid-1970). These were compared with the AP (accounting period) accounts-thirteen x four weekly accounting periods during the year, to reflect changes in performance and to revisit the expected yearly results which would have a bearing on the amount paid for interim and final declared dividends. This would be phased, or split, across months in the financial year and would show the cumulative result to the end of each four weekly accounting periods after which, a financial report would normally be run.

A revision of both the five-year forecast and the short-term forecast should have indicated the increasing pressures on the cash resources of the group.

Management Accounts: financial reports produced for Directors and managers and would usually contain a profit and loss report and balance sheet. These are similar to year end accounts but less formal and can be tailored to requirements. These would show the cumulative actual results – not forecast results, to the end of each of the accounting periods to monitor performance against the annual, phased profit plan.

The Product Profit Plan: Basically, this is a Profit equals Price minus Cost calculation used to determine the initial viability of a product and show the profitability of a product over its entire life span. Its purpose is to show the amounts and timings of revenue, costs and profit or loss for any project. Profit plans were subject to a high amount of uncertainty given the projection might be 10-15 years ahead, together with the general uncertainty associated with the aero engine industry.

At the July main board meeting in 1967, the board were presented with a product profit plan for sales between 1970-1979 for the RB211-06 (the earlier development stage engine of the RB211-22), which would be initially offered to both Lockheed and MDC, with assumptions that there would also be a higher thrust version as well as engines sales for the RB207 twin-engine aircraft. The RB207, also included in the profit plan would be offered for the European Airbus. Predicted were sales of 3,289 engines over 10 years (1970-1979) -510 for the L-1011 and 500 twin with 30% spare engine requirement. The forecast was predicated on the consensus of opinion at the time from airlines and the aircraft supply industry, that there would be only one engine manufacturer and one airframe manufacturer. However, this would not turn out to be the case.

In this plan, the launch cost estimate of the RB211-06 engine, initial application was £55.7 million. With the initial rating for the twin-engine application plus a higher thrust version and continuing development, the total launch cost would be £91.4 million for which HMG would be asked to contribute 70% of the costs.

The forecast for engine sales between 1970-1979, on which the £91.4 million estimated total launch costs were based, including for twin and higher thrust versions, was presented to the Ministry on the 29th of June 1967 and subsequently used as the basis going forward for further product profit plans up to the signature of the contract the following March.

Over the whole life of the project, it was estimated that the net profit, based on the total launch costs estimate of £91.4 million, would be £273.8 million.

However, a lower launch cost of £83.1 million was used in order to arrive at the profit forecast. and was subsequently used as the basis going forward for further product profit plans again, up to the signature of the contract. This figure assumed efficiency savings would be made by the engineering department which had been set out in the five-year plan. In the event, due to a number of factors, including poor cost control, these efficiencies were never fully achieved.

On the basis of the product profit plan, the board gave authority to make final offers to Lockheed and MDC.

In December 1968, almost a year and a half later, a new RB211 product profit plan was produced by DED. This showed the estimated sales had contracted from 3,289 to 2,515 engines and only 720 of these were for the basic L-1011. A further plan was produced in July 1970 and projected reduced sales figures of 1,217 engines for 338 aircraft and 150 engine kits projected for the industrial and marine market.

Rather than one airframe and one engine manufacturer in the market, the TriStar and DC-10 were competitors even on the same flight routes and Rolls-Royce were competing with GE to supply the engines. In the end only 249 TriStars were eventually sold.

The 'Metcalfe Report' which would be presented in June 1969, extrapolated figures for the RB211-22 alone from the plan incorporating the whole project which included the stretched – higher thrust versions, and showed that rather than a profit, the RB211-22 was showing a loss.

<u>The AP Accounts</u> were used as a control of DED's financial affairs on a day-to-day basis. The first comparison of actual results with the phased annual profit plan in any year would normally be prepared at the end of the third accounting period. Due to the time taken to produce the year end accounts, the results normally available by mid-April, would not be reported to the divisional board until the May board meeting, five months into the new year.

Had matters been stable this may not have been so much of a problem but given the increasing costs of the project with the decreasing cash flow, the information did not reflect the situation which had moved on by the time it was presented. The AP accounts would be presented to the divisional board and although containing a wealth of information-some 70 pages, daunting for anyone to absorb, particularly in a short space of time from its presentation; the Inspectors, MacCrindle & Godfrey in their investigation under S165(a)(i) of the Companies Act 1948 into the affairs of Rolls-Royce, in their Report to Government, Ibid, (1973) (pg. *89)*, concluded that the way the information was presented did not 'enable the divisional board members to properly appraise the performance of the division and to ensure the best use of the division's resources'. The Inspectors also criticised the lack of analysis, including that contained within 12 of the 70 pages, in a Report by the divisional controller. Although there was plenty of information available it was not in a format that was clear enough to enable it to be easily understood.

POSSIBLE CONTRIBUTING FACTORS LEADING UP TO THE 1971 ROLLS-ROYCE LTD. BANKRUPTCY 1961 TO 1971

Change in Accountancy policy: In 1961, Rolls-Royce changed its accountancy policy relating to capitalisation of its R&D. This change meant that the then current policy of writing off, of some of the research and development expenditure against profits in the year the money was actually spent, would change to the R&D expenditure being carried forward over a number of years. By doing this Rolls-Royce could continue to raise funds by showing adequate profits. The change made in 1961 was adopted for all accounting periods going forward up to the RB207 and RB211 R&D expenditure in 1967, when a different basis was adopted for these two engines. As explained in notes within the company accounts. . .

> *"during the early years of the development and production of the RB207 and RB211 engines for which orders had not been received at the year end and would be likely to distort the profit figures because of the magnitude of the expenditure and the uncertainty of the timing of the receipt of future orders.*
>
> *Therefore, in respect of these engines it was intended to adopt the practice of amortising net launching costs over estimated further sales, and a sum of £532,000 representing such expenditure has been carried forward as an asset at 31st December 1967".*
>
> Ibid. *MacCrindle R.A. QC.,& Godfrey P., FCA (1973) (pgs.51-52).*

Regarding amortisation;

> *". . . This was a perfectly acceptable accounting practice and is constantly used throughout the world by companies engaged in research involving high expenditure and long development lead times.*
> *. . . Over the years 1961-1969, Rolls-Royce decided to adopt a gradual method of writing off its research expenditure. It did not make any difference to the actual level of expenditure – the money was still spent whichever method of accounting was used – but it made a difference to some people's perception of the balance sheet. Over this 9-year period the change meant that reported profits were £11.7M higher than if previous methods had been used".*
>
> **Ibid Pugh Peter,** (2001) *(pg. 139)*.

Decline in profits equals increased borrowing:

In the aero engine industry R&D investment can take years to recoup and can cause a continual drain on financial reserves, meaning that dividend payments and growth are financed either by borrowings or equity stock which in turn has a further impact on the profit and loss account due to interest payments, or extra dividends being announced. R&D spend was a continual drain on the company's resources.

In 1961 a profit loss would have been reported but for the change in accountancy policy. In 1962, 1967, 1968, & 1969 the profit would not have been enough to cover dividends paid.

Compared to other industries, aerospace has always incurred a higher than average dependency on borrowings to fund business over several years. In 1961, Rolls-Royce's dependence on long-term and short-term borrowings increased from 22% to 46%. By December of 1966, following the takeover of Bristol Siddeley Engines, 49 % of the group's resources after deduction of intangible assets, such as Patents, were financed by borrowings. As reported to the main board in May 1967, the company's bankers had agreed to additional overdraft facilities for 6 months on the basis of the funding operation by Government. This increased the overdraft and acceptance credit facilities to £71 million, but even these facilities would be exceeded by August 1967, according to the group treasurer's report presented at the same meeting. Subsequently £17 million was raised by the rights issue in September 1967 and the sale of shares in Westland, as part of the purchase of Bristol Siddeley Engines, amounting to £2 million. At the main board meeting toward the end of 1967, concern was expressed that despite the above, borrowings were still too high.

In January 1968, the main board finance committee prepared a paper to present to the main board warning of lower profits going forward and an increased reliance on short-term borrowing. Amongst other matters it went on to state that over £23 million would be required to be earned before tax to pay Debenture interest, interest on short-term borrowing of over £60 million and a dividend of 11% with a debt ratio of 1.4 covered- meaning for every pound of shareholders equity the company owed £1.40 to creditors.

In June 1968 the Company announced that it was raising £21 million by way of a rights issue and £10 million by way of a Debenture placing. The full proceeds would be received in November and would reduce borrowings as a percentage against capital. However, affecting the planned capital requirements by the end of 1968 were: the bringing forward by two years the development of the L1011 (long range version), which would increase the engine R&D spend some £40 million over seven years, which was without government commitment of financial support and, as mentioned previously, the market estimate for sales having reduced. Profits estimated at £39.4 million were not in 1968 forecast to occur before 1974. Prior to the above mentioned becoming affecting factors both were included in the original Product Profit Plan, producing more favourable figures than for the RB211 engine alone. When extrapolated, the figures for the RB211 alone would show the 'adverse variance' reported to the main board in June 1969, by the RB211 Programme Director, Mr. T. Metcalfe.

Conversely in 1968, Rolls-Royce reported its largest pre-tax profit of £15.9 million, with a slight improvement in its operating margin, achieving a record turnover of £319.6 million. However, after 1968 there would be a declining trend in the operating margin and turnover would fall by some £20.3 million in 1969, which was an altogether poor year financially for the company. Pretax profits were £6.4 million compared to the 1969 profit plan of £15.3 million and having achieved £15.9 million in the previous year. Short-term borrowing had increased by £7.6 million, DED operations accounting mostly for this and whose profit before interest payable had fallen by £9.3 million. DED's cash needs through to 1973 had increased by £49.7 million compared to the 1969-73 forecast of November 1968.

Revised five-year forecasts called for by the main board in March 1970, were presented at the meeting in April 1970. It showed a further decline in profits in 1970 and 1971 compared with the October 1969 forecast but profits were estimated to increase in 1972 as the decision had been taken not to proceed with the RB211-50 (stretched version). Whilst the forecast implied an increase in liquidity problems with increased borrowing within those two years, following on from those, it was expected to quickly recover. It was decided to go ahead with the loan from the IRC of £10 million.

Mr. Trevan-Hawke had written to the IRC in April 1970, on the advice of the Rolls-Royce Bankers, Lazards, who were concerned as to the unsatisfactory financial position of the company and presented an estimated five-year plan forecast for 1970-74 referred to as 'Issue B' inviting a comparison with the five-year forecast previously sent to the IRC in 1969.

A revised product profit plan was produced following revised launch costs estimates made known in July. This showed potential project losses amounting to £98.9 million. A comparison with the Issue B plan with a revised forecast based on the new product profit plan was presented to the mainboard in a report by Mr. Trevan-Hawke in July 1970. This attempted to set out the effect of the project on the company's profits and finance required. Compared with the 'Issue B' forecast sent to the IRC the report stated that the effect on the estimated profit after savings and reductions would be 'distinctly modest with none at all until 1973'. It showed only a 3.8% profit after tax in 1970, decreasing again in 1971 to 0.5% and only rising slightly in 1972 to 1.3% before anticipated rises to 8.6% in 1973 and 11.9% in 1974.

The corresponding effect on borrowing facilities required from estimates produced would need an increase in overdraft facilities by £5 million in 1970 of the then current overdraft limit of £70 million. The figures estimated overdraft limits increasing substantially in 1971 to £90 million to through to a peak in 1972 of £105 million, but even though estimated to then decrease through to 1974, it was still more that the then agreed overdraft limit by between £5m and-£10 million. By 14th October 1970 in a revised group five-year forecast, having included a margin of £20 million for contingencies, the peak cash requirement would be stated at £119.5 million some £49.5 million over the overdraft facility of £70 million.

		1970/74 Plan 'Issue B' presented to IRC			Revised Forecast July 1970 presented to main board finance committee		
Year	Profit after Tax	Dividend Rate	Overdraft	Profit after Tax	Dividend Rate	Overdraft	
	£M	%	£M	£M	%	Estimated £M	Limit £M
1970	5.5	8	62.5	3.8	5	67.4	75
1971	3.8	8	67.5	0.5	5	84.1	90
1972	10.9	13	70.6	1.3	5	93.3	105
1973	18.0	22 1/2	52.0	8.6	5	67.1	75
1974	27.4	25	49.1	11.9	12	69.8	80
Totals	65.6	51.3		26.1	21.3		

Reproduced from Ibid., (*MacCrindle RA & Godfrey P. FCA (1973) pg237*)

Sort out the problems ahead of development:

Anyone working in aero engine manufacturing will know the design and development of an engine often takes several years and is a complex affair. The preliminary design functions are definition, concept and detail. The preliminary stages lead to mechanical design. The design is then tested using development engines which are specifically built for this purpose. During this phase any design flaws are revealed and corrected. In addition, the engine performance is steadily improved to the specification required by the customer. The specification can change, as it did with the RB211, in line with any changes required by the customer and any problems with the airframe during its development. Adjustments with the engine design may result in overcoming those problems. Any design change has a knock-on effect on the development, experimental, manufacturing and production departments. Prototype engines are built following the development phase for flight testing which takes place before final certification for civil aircraft use. Throughout all these stages, the engine can be altered or improved, and modifications can be required throughout its engine life. However, improvements in one area may be at the expense of another.

The design of the RB211 engine was something quite new to Rolls-Royce incorporating new and major technological advances: the three-shaft concept, use of composite materials such as Hyfil and the use of an annular combustion chamber in a high-pressure engine. The size of the engine and thrust required and subsequently offered was greater than anything that had been achieved before.

All of this innovation was more likely than not to incur overspending in R&D. It had been proposed to design and develop the engine for entry into service in 1971, giving four years for completion of the development. However, the development of the long-range, L-1011 was brought forward from the second to the first year.

The company's experience of this type of engine was limited to the RB178 demonstrator engine, (the forerunner of the RB211) which Rolls-Royce started to build in 1965 under a demonstrator contract awarded by HMG and work being developed on the Trent engine. In 1966, the RB178 engine programme exposed several problems, after having only run for a few hours, particularly with the combustion chamber and High Pressure (HP) turbine. The programme was suspended at the end of 1966 due to an escalation of project cost overruns of £218,000 in excess of the approved budget, with no airframe application identified and a reported lack of management interest. Sir David Huddie is reported to have described the cancellation as "one of our great mistakes" and it was later viewed that many of the early RB211 development problems would have been overcome more quickly had it been extended further.

The only other experience was the development work on the concept of the three-shaft engine of the Trent in July 1966, which ran for the first time approximately 3 months before the signing of the RB211 contract before it was cancelled in July 1968, which provided confidence in the three-shaft concept.

A high temperature demonstrator unit was opened by DED, to develop engineering knowledge in this area and would have assisted with the RB211-22 turbine blade cooling problem. The unit initially ran in February 1969 and closed in early 1970 following the company rejecting the governments offer to continue to fund 75% of the running costs up to a limit of £300,000 but not all of yearly cost of £400,000 which Rolls-Royce had asked for.

The production release process meant that designs for tooling and equipment for production components were issued for the RB211-22 programme as early as September 1968. This was despite the early development engine, the RB211-06, having made its first planned run in August of that year, highlighting there would need to be some re-design required to meet specifications on fuel consumption and thrust, within the limits of the turbine inlet temperature. Rig tests showed that some of the deficiencies were due to inadequate basic component performance. Many of the early release of designs proved worthless due to the number of specifications and design changes required to fulfil the contract, with some of the designs released not having been tested. This pressure of design and redesign within the timescales given, affected all areas of engineering, experimental, manufacturing and production, resulting in:-components being ordered then cancelled as the design had moved on, development drawings not recognised by the configuration control system, the engine having to be stripped because it contained the wrong part number or for the component to fail due to incompatibility of drawings between schemes.

In a paper presented by DED contracts director Joe Clayton entitled Contract Lessons dated 27th April 1970, believed . . . "more emphasis should be placed on. . . . independent development of critical components and on getting them quickly to test" . . .. He thought the vicious cycle of not supporting an idea for development, even if there was no requirement, should be broken.

Ensure sufficient resources are available to deliver the project

Engineers were spread across projects already underway involving the Spey for the Phantom aircraft as well as projects for the Adour and Trent. Senior engineers whose talents were needed in Derby were also used in the sales campaign, providing technical liaison, many having to frequently visit US airframe companies, MDC and Lockheed to discuss specifications. In September

1968, John Bush, Director of Engineering reported that he urgently needed experienced and talented designers. The programme needed the best and most senior designers to concentrate exclusively on the RB211 engine. Whilst later in October 1969, Sir David Huddie believed that an arbitrary reduction in manpower was required to reduce operating costs, the 1969 published accounts demonstrate a loss of employees within the group as a whole by 1,230 within the year from December 1968 to 1969, whereas the number of employees in the DED engineering department rose by nearly 300 reflecting the engineering requirement to cope with the difficulties being encountered with the RB211 engine.

The engineers were reported by Tom Metcalfe to Sir David Huddie in November 1969 as "working flat out . . .but not making positive technical progress anything like as fast enough" and although he acknowledged the "first class and experienced down-to-earth development men" in Messrs. Bush, Hare and Mayall – "they needed more of them".

It also reflects the problems of reduced development time. For example, bringing forward the requirement for increased thrust in the RB211-22 to 40,600lb to the initial year of service from the second, would mean the timescale for completing the development of the engine had likewise reduced. Additionally, the F.A.A Stage 2 noise rating would also have to be brought forward, causing extra pressures on the engineers particularly if there were any further specification changes, which when the contract was signed, large areas of which remained unresolved.

What price goodwill

The acquisition of Bristol Siddeley Engines (BSE) in 1966, at a price of £63 million (£21.9 million of which was goodwill) was seen as a large drain on the vital cash resources of the company, particularly at that time, given the impending investment requirements of the new 3 shaft engine RB211 project. However, as previously referred to in 1966, BSE was a competitor of Rolls-Royce and was threatening to enter into a licence agreement with US engine maker Pratt & Whitney to manufacture its JT9D engine, so a takeover of BSE by Rolls-Royce was seen as a strategic move to eliminate a competitor from the market. Air Commodore Francis Rodwell Banks, Rolls-Royce Adviser and a former Air Ministry Director, of Engine Research and Development, held strong views that this had been an unnecessary move leaving Rolls-Royce with considerable overheads and a loss of a competitor to 'keep them on their toes' (December 1971).

Mind your net worth

Net worth is the value of all the non-financial and financial assets owned by an institution (or individual) minus the value of all its outstanding liabilities. It appeared, neither the DED nor main board had fully evaluated the risk involved of any overspend on the estimated cost of launching the engine which would fall entirely on Rolls-Royce and the escalating effect on its net worth. The main factor that could have been anticipated in the design of an advanced project was the technological demands due to specification changes, which was what occurred. Additionally, staff requirements, wage demands, inflation, deflation and currency fluctuations can all have a bearing and mostly did.

From the initial estimated £30.9 million cost of the dual programme and continuing development of the higher thrust version against the estimated net worth, (excluding intangible assets) in December 1967 of £94.2 million, the company would commit nearly one third of its net worh to one project.

The financial review which had been requested by the July 1967 main board meeting was presented to the board in September and took the format of a revised five-year forecast for 1967-1971, based again on the above programme. In this review the net worth of the company was forecast to increase from £94.2million in December 1967 to £122.7 million (after deducting intangible assets).

The initial launch costs borne by Rolls-Royce had increased from the £30.9 million in September 1967 to £67.97 million for the RB211-22, sold to Lockheed in 1968. This sum takes into account HMG launch aid at the maximum £45.75 million, plus an additional £1.38 million (which would be added when launch aid terms were finally agreed with HMG in August 1969, transferred from the RB207 to the RB211-22 programme), accounting for the total of £67.97 million of investment indicated.

In fact, the 1967 published accounts stated the net worth figure (excluding intangible assets), was £90.8 million. By this sum, together with the £21.7 million in equity raised in 1968, the company had on this current forecast of costs, risked 60% of its net worth in the RB211-22 programme from the 30% previously estimated for the RB211-06 programme.

The estimated launch costs of the RB211-22 engine provided to MinTech in September 1968 by Sir David Huddie, was for the RB211 engine alone, £74.9 million, with launch costs applicable to the thrust reverser and pods taking this to £82.20 million and with continuing development and replacement tooling added, the total was £89.3 million.

The engine cost would increase to £84.1 million and with the added costs the total would take it to £99.6 million by June the next year. This would escalate continually and by the time the receiver was appointed the launch costs would be around £195.5 million which was 2.19 times the figure submitted to MinTech in 1968. With the need to lean on its equity capital and reducing reserves coupled with increased borrowing, the financial risk on the company's net worth was ever increasing.

Engine unbuilt-untested

At the time of the signing of the RB211 contract with Lockheed in 1968, large areas of the engine design and contract specification were still unresolved, hence the reference often used that the parties had signed up to a 'paper engine', i.e., an unbuilt and untested engine. A final figure for the launch costs of the project could not be completed and submitted to HMG for over a year, so delaying confirmation of launch aid.

Pitfall of accepting late contract changes–don't lose sight of the cost

Following soon after the signing of the RB211 contract, in March 1968, Lockheed asked that a change be made to the mounting of the engine which involved moving it from the rear of the engine core to the top of the fan casing. This was agreed by Rolls-Royce at no cost to Lockheed. The objective was to save aircraft weight at the expense of engine weight but in the event, the engine weight increase was underestimated and making a requisite increase in thrust necessary. Later acknowledged by Sir Denning Pearson, that Rolls-Royce agreed to this change too easily given that the programme was well underway.

During the course of negotiations, the specification changed, amongst other things, in weight. noise guarantees, noise attenuation materials and thrust. With the first running of the engine planned for August 1968, the design and development work were already up and running and so quickly were the costs!

The table on pages 149-151 shows the main price quotations to Lockheed for the RB211 basic & also for installed engines which included accessories and deferred payment terms, for deliveries in 1971, with engine designations and thrust variances. (McDonnell Douglas (MDC) prices and thrust offered included where known).

Date	Engine Designation	Basic Engine Price	US dollars@ $2.80 to the £Sterling	Installed Engine price & accessories including deferred payment terms unless otherwise stated. £ sterling	US dollars@ $2.80 to the £Sterling	Thrust offered	Temp.
23.06.67	RB211-06	£179,500	$502,600	£193,178	$540,900	34,250lb. 33,260 lb.	84° F 90° F
13.07.67	RB211-06	£182,500	$511,000	£196,178) (see Chronology Chapter 6.(para 6))	$549,200 (see Chronology Chapter 6.(para 6))	34,250lb. 33,260lb.	84° F 90° F
20.07.67	RB211-06 Offered to MDC	Prices quoted as per those previously advised to Lockheed					
20.10.67	RB211-10 offered to MDC			£198,267	$555,150	35,400lb.	84°F
01.11.67	RB211-14	£189,664	$531,060	£199,964 deferred payment terms now included in engine price	$559,900	35,750lb. 34,740lb.	84° F 90° F
		Devaluation of Sterling	Conversion to US Dollars now @$2.40 to the £Sterling		US Dollars@ $2.40 to the £Sterling		

01.12.67	RB211-15	£221,875	$532,500	£232,292 all previous offers to Lockheed withdrawn. (Previous quotes would take reduction in exchange rate into account)	$557,500	35,750lb. 34,740lb.	84°F plus specification changes 90°F
06.12.67	RB211-10			Reduction of 5% offered to MDC £188,353	$527,393	35,400lb. Increased thrust requested	84°F
05.01.68	RB211-15 With provision for handshake Thrust of 5%	£225,000 based on previous figure quoted for RB211-15 & incorp. $7,500	$540,000	£235,417	$565,000	35,750lb. 34,740lb.	84°F 90°F
05.01.68	RB211-17 quoted to MDC					38,000lb.	84°F
22.01.68	RB211-18 offered to MDC					38,000lb.	84°F
14.02.68	Price reduced by 3% to Lockheed for all firm orders placed before 30.09.1968	The price this was based on is not referenced.					
01.03.68	RB211-15 (Including Handshake thrust	£221,875 Incorp. $7,500	$532,500	£235,417	$565,000	35,750lb. 34,740lb.	84°F 90°F

01.03.68	RB211-18	£228,125	$547,500	£238,542	$572,500	38,000lb. 37,000lb.	84°F 90°F
06.03.68	RB211-18(stage 2) to MDC	£227,083	$545,000			40,600lb.	84°F
06.03.68	RB211-18(stage2)to Lockheed	£230,208	$552,500	£240,625	$577,500	40,600lb. 39,576lb.	84°F 90°F
14.03.68	RB211-22 (stage1) with shorter turbine life to MDC & Lockheed	£225,000	$540,000	£235,417	$565,000	40,600lb. 39,576lb.	84°F 90° F
14.03.68	RB211-22 (stage2) to MDC & Lockheed	£227,083	$545,000	£237,500	$570,000	40,600lb. 39,576lb.	84°F 90° F

Author's table with figures reproduced from the Inspectors report Ibid (*MacCrindle R.A,QC & Godfrey P. FCA, (1973)*. (The Report quotes deferred payment terms for the 13.07.67 offer as being $13,000; £4,678, the same amount in Sterling as the deferred payment term relating to the firm offer made the previous month of $13,100 ie £4,678.).

CHAPTER 6

CHRONOLOGY OF RB211 KEY EVENTS AND ESCALATING COSTS 1967-1971

The below may assist in providing a better understanding and perspective of some of the main RB211 background events taking place from the early development of the engine, the signing of the contract in March 1968 and the worsening effects on the company's balance sheet, leading up to the bankruptcy in February 1971. Information compiled in this chapter comes from several sources, particularly Ibid *Hayward K. (1983); Pugh P. (2001) and Ruffles P. (2014)* but predominantly from the 'Report to Government', Ibid *MacCrindle R.A., QC & Godfrey P., FCA,(1973),* and full acknowledgement is given to all sources.

To be borne in mind, only costs attributable to engine launch costs would qualify for launch aid from HMG of up to a maximum of 70%. In addition, the figures noted in the chronology can refer to development costs; production cost estimates; preliminary engine costs and prices; Installed engine costs, which included a basic engine plus accessories (for deliveries in 1971 and before 1st November 1967, deferred payment terms). Also refer to the schedule in the previous chapter. A propulsion system consists of an installed engine plus the relative power plant items and pod. A 'shipset' includes three installed engines, needing three sets of power plant items and a set of three pods being two wing fitments and one tail fitment. The engine prices are stated, as were offered to Lockheed and McDonnell Douglas (MDC), in US dollars with the equivalent price in sterling applicable at that time period. Any production costs are shown in sterling.

Even though not all the different types of financial information are recited, it demonstrates the almost immediate, escalating problems Rolls-Royce faced.

1. **In February 1967**, a main board sub-committee considered amongst other matters, profit plans for the RB207 and RB211-06 engines as a dual programme. It agreed firm offers could be made for these engines at specific prices for the European airbus and the Lockheed L-1011 aircraft. Those involved could make 'modest changes in prices

during the competitive phase, provided it did not significantly alter the commercial facts of the proposition'.

2. **On May 26th, 1967**, Sir David Huddie wrote to the Ministry of Technology (MinTech) explaining that US studies were showing if a three-engine airbus was proceeded with, then an engine generating a thrust of 33,000lb. would be required. The appropriate version of this would be the RB211 engine. Launch costs were put forward, in principle agreed, with conditions for either the RB207 or RB211 engine applications, to a maximum of £38 million or £40 million in certain circumstances.
3. **In June 1967**, Lockheed as a potential supplier to American Airlines, confirmed to Rolls-Royce they were looking at a Trijet, not a twin-engine aircraft previously quoted for and were offered the RB211-06 engine with 33,260lb. thrust on a 90°F day, subject to conditions and main board approval; the RB207 development programme was scaled down.
4. **On June 23rd, 1967**, Rolls-Royce offered to Lockheed, the RB211-06 engine, uninstalled plus accessories, as a firm price of $527,800 (£188,500) @ $2.80 to the £ Sterling (excluding deferred payment terms which would incur a further increase of $13,100 to $540,900 (£193,178)) on its usual terms being, 15% payable before delivery with balance on delivery. This sum excludes the cost of power plant items such as thrust reversers, which were made on an indicated price only basis.
5. **On June 29th, 1967**, projected sales figures over a ten-year period were submitted to MinTech and linked to HMG providing the launch aid of 70% of the estimated launch costs of £60.25 million for the RB211-06. Sales of engines forecasted were for 1,989 engines in the US aircraft, a further 1,300 engines for twin-engine aircraft (the figures included 30% spare engine recommendation) and included forecasts for an increased thrust version for which development costs were estimated at £20 million more. However, MinTech were told that launch aid towards the development of the increased thrust version was not requested at that time. These figures were the basis of the product profit plans prior to the signing of the contract.
6. **On July 13th, 1967**, a price increase of $8,400 (£3,000) per engine was notified to Lockheed. This followed on from the firm price quoted on the 23.06.67 and before the reply from MinTech the next day. This increase effectively equaled the increase in margin for the recovery of launch costs indicated by MinTech on the 14th July from £9,000 to £12,000, for each RB211 engine sold. The Ministry of Technology having modified the original proposal by Rolls-Royce to them.

7. **On July 14th, 1967,** MinTech replied to Sir David Huddie confirming support from HMG was conditional upon receiving a firm order from Lockheed 'leading to an adequate scale of production', making it more economic. It indicated support for launch costs would be limited to £40 million, 2.15 million less than Rolls-Royce had asked for.
8. **On July 18th, 1967,** The Rolls-Royce main board approved the product profit plan for the RB211-06 and RB207 programmes with the forecasted sales as earlier outlined:-3,289 engines to be used in 510, L-1011 aircraft and 500 twin engine aircraft including provision for 30% spare engine sales and an assumption that there would be a higher thrust version of the RB211-06.

   The RB211-06 development programme was launched. The board emphasised the importance of a financial review to show the effect on the company of supporting both programmes, or just the RB211 alone.
9. **On September 19th, 1967,** the main board were presented with the five-year forecast, 1967-1971 based on the dual programme who agreed the programme was financially feasible *only* if HMG provided launch aid on the plans which had been submitted to MinTech. Although the forecast for 1968-1971 suggested that profits would absorb the costs of the group's entire R&D of £60 million and increase retained earnings (the amount of profit a company has left over after paying all its direct costs, indirect costs, income taxes and its dividends to shareholders) of £13.8 million. However, the board were also told that short term borrowing was expected to increase between 1968-1971 by £18 million if a proposed loan stock issue did not take place.
10. **On November 1st, 1967,** Rolls-Royce were keeping MinTech informed of developments and in a meeting raised the issue of criticism from the US regarding the RB211 having no prior flight experience. Consequently, the company requested use of a RAF VC.10 aircraft as a flying test bed but were told further discussions would take place if the contract were obtained. Details and cost of the RB211-14 installed engine were provided to Lockheed. The price had increased $19,000 (£6,786) to $559,900, still at $2.80 to the £ Sterling (£199,964) from that quoted on the 23rd of June 1967and $10,700 (£3,821) more than on the 13th of July 1967. Going forward, deferred payment terms would be included in the basic engine price. In addition, the basic engine price plus accessories would now form the installed engine price.
11. **November 18th, 1967,** Sterling was devalued, falling by approximately 14%. Rolls-Royce felt able to quote new dollar prices.
12. **On December 1st, 1967,** All previous offers made to Lockheed were withdrawn. The RB211-15, with some changes to the specification

and for engines delivered in 1971, but with the same engine thrust as previously offered, was then quoted at $557,500 (£232,292),with an exchange rate of $2.40 to the £ Sterling for an installed engine. Lockheed were informed that any previously quoted prices would have been reduced by 5% to take into account, the change in exchange rate. During December 1967, both Lockheed and MDC made requests for additional thrust. In addition, the various alternative quotes for acoustics, for example, different price options for limited or extended use of acoustic lining in the fan outlet duct, were withdrawn and included in a single price quotation.

13. **On December 6th, 1967,** a 5% discount to the price previously quoted, was offered to MDC, which accounted for the devaluation in Sterling. However, due to Lockheed having made changes to their requirements, this was not made to Lockheed. Toward the end of December, some board members were expressing concern regarding the continued high level of borrowing, despite the £17 million raised from the Rights Issue and £2 million from the sale of Westlands, which were aired at the main board meeting in December.
14. **On January 5th, 1968,** Lockheed wanted to know if 5% extra thrust could be supplied if required to meet aircraft guarantees. Lockheed were informed it could but at extra cost of $7,500, plus a further $10,000 if Lockheed were required to operate at the higher rating requiring development work and certification. This was termed a 'handshake thrust'. Rolls-Royce also informed Lockheed (and MDC) that a previous stipulation of a price increase of $9,520, should a titanium fan blade be preferred over Hyfil, would be withdrawn, confident that Hyfil would be chosen and in order that customers could be made an unqualified financial offer. The RB211-15(c) was offered at $565,000 (£235,417) to Lockheed with the same thrust as the RB211-15, ie 35,750 lb. at a day time temperature of 84° but with the provision for a 'handshake thrust' increase of 5%. The same day Rolls-Royce offered the RB211-17 to MDC, but information on the thrust and price have not been identified.
15. **On January 30th, 1968,** a paper prepared by the finance committee which included revised financial estimates was presented to the main board. The forecast included estimated cash requirements of £20 to £30 million by the end of 1971 and this was in excess of the potential for short-term borrowing of £18 million indicated to the board in September 1967; and profits after tax some 23% lower than contained in the five-year profit plan on which it had based its decision to go ahead with the RB211/207 projects.

16. **Throughout January and February 1968**, reductions in prices were made; additionally, specification changes were requested and deferred terms being offered. Prices quoted for acoustic materials reduced within a short space of time, from an initial price indication of $49,000 (£20,417) in December to $54,000 (£22,500) in January and from then on reducing to $25,000 (£10,417); a decision had been taken to use a plastic material, in place of the more costly felt material, all to retain a price advantage and all imposing additional strain on company resources, adding to concerns.
17. **On February 14th, 1968**, the price per engine was reduced to Lockheed, by a further 3% for all firm orders placed before September 30th, 1968, in response to a reduction in price offered by GE. The same offer was made to airline customers with certain provisions.
18. **On February 19th, 1968**, on reviewing the January paper from the finance committee Sir Denning Pearson believed if Rolls-Royce could 'secure an order of this magnitude, finance would be forthcoming somehow'.
19. **On March 1st, 1968**, revised quotations for the RB211-15 with various levels of 'handshake' thrust were sent to Lockheed which incorporated in the standard price, the $7,500 previously notified on January 5th, for the 'handshake' thrust. At the same time, the price for the RB211-18 was also offered at $572,500 for an installed engine (£238,542) $7,500 more than the initial price of the RB211-15 offered in January, offering a thrust of 38,000lb. on a day temperature of 84°F and 37,000lb on a day temperature of 90 ^{0}F exceeding that of the RB211-15.
20. **On March 6th, 1968**, following revised thrust changes and following American Airlines' order for the DC10, the remaining airlines; Eastern, Trans-World Airlines, Delta Air Lines and Air Canada, asked for increased thrust to be available for initial service. It was agreed that the RB211-18 thrust increase to 40,600lb. on an 84 ^{0} day, would take place within the first year of service. It had been intended to develop the thrust to this level during the second year although it was accepted that the life of the initial high pressure turbine blade would be limited to only 5,000 hours. Therefore, the thrust requirement had been increased but the timescale for completing development had been reduced, increasing pressure on the engineers. The FAA noise regulations-'stage 2 rating' would also be brought forward. The price offered for the RB211-18 stage 2 basic engine was $552,500 (£230,208) and for the installed engine, $577,500 (£240,625). The commercial department in Derby suggested the basic engine price of $555,000 should be quoted for the stage 2 rating (an increase again of $7,500 over that previously quoted to

Lockheed for the RB211-18- basic engine price of $547,500; (£228,125). Sir David Huddie notified constructors and the airlines on 6th March of an engine price increase of only $5,000 per engine for the stage 2 rating. This would have made the basic engine price to Lockheed, $552,500 and to MDC $545,000, creating a price discrepancy of $7,500 between that offered to Lockheed and the price offered to MDC. Pointed out by the commercial department in Derby it was picked up by the sales project group in America who said that unless the engine design differed significantly then the basic engine price to both Lockheed and MDC should be, for the Stage 1 rating (which would have a shortened high pressure turbine blade life), $540,000 (£225,000) and engines to full stage 2 rating, $545,000 (£227,083); making this respectively,$7500 & $2500 less than quoted for the RB211-18 at 37,000lb. thrust.

Following this, the engine designation would now be RB211-22.

21. **On March 7th, 1968**, a report of this date was presented to the DED board by the DED financial controller, Jim Rigg, expressing disquiet over the 1968 profit plan as results were falling well short of the Five-Year forecast on which the board had based its decision to go ahead with the RB211 project.
22. **On March 14th, 1968**, formal offer letters were made to Lockheed and MDC confirming the price of the RB211-22 (stage 1) for an installed engine as $565,000 (£235,417) and the stage 2 engine as $570,000 (£237,500), both at a thrust of 40,600lb. Over a year would then elapse before the specifications were resolved to the point where a revised estimate of the launch costs could be submitted to MinTech.
23. **On March 29th, 1968**, the RB211-22 engine contract was signed between Rolls-Royce and Lockheed Corporation and an order was placed for 150 ship sets of the RB211-22 propulsion system. This was the largest engine Rolls-Royce had ever built; the three-shaft engine design was unique as was the proposed innovative use of the carbon fibre composite material, Hyfil and having an annular combustion chamber in a high-pressure engine. Soon after the signing of the contract Lockheed asked for a change to the mounting of the engine. Rolls-Royce agreed to the change without cost to Lockheed. Eastern airlines had ordered 50 L-1011 and TWA 44. A new five-year forecast was prepared.
24. **On May 21st, 1968,** the new five-year financial forecast, 1968-1972, the draft of which had been discussed with Lazards the Investment Bank, on May 2nd, 1968, had shown a potential further funding requirement of between £25 million and £35 million.

25. **On June 28th, 1968,** the company announced that it was raising circa £20 million by way of a rights issue and £10 million by way of a debenture placing.
26. **In July 1968, RB211-22** production costs estimate alone, based on the design at the contract date, of an installed engine was £153,655 excluding development, administrative, selling and financing expenses. The selling price of an installed engine to Lockheed under the contract for deliveries in 1971 was $570,000 (£237,500.00) (see paragraph 22 above) however, for a period, a temporary price was offered of $552,900 (£230,375), for all firm orders placed before 30th September1968. On the basis of the temporary figure, it was estimated the gross margin from the sale of each engine in 1971 would be £76,720,approximately 33.3%.
27. **On August 1st, 1968,** the Head of Staff, Design, Don McLean, wrote to the DED Director of Engineering, Ernest Eltis, having critically reviewed the engine identifying twenty-six components which needed urgent design input.
28. **During September 1968,** a release of extensive and detailed drawings was given to Production, much of which proved worthless due to further, later design changes, putting additional pressures on the design team. The production release process had started to build up stocks of components whilst engine development was still in progress, to avoid delay.
29. **On September 2nd, 1968,** David Huddie, wrote to the MinTech with estimated total launch costs of £89.3 million and requested a 70% contribution toward the engine launch costs of £74.9 million, being £52.4 million. The costs having increased from £55.7 million for the RB211-06 due to changes in the design specification. The Ministry rejected this due to not being able to confirm the precise costs.
30. **In November 1968,** development of the long-range version of the L1011-500 would be brought forward two years earlier than planned. The RB211-50 (RB211-524), a stretched or thrust growth version of RB211-22B was to be developed for this.
31. **In December 1968,** a new RB211 product profit plan was produced by DED. The estimated market had reduced from the total 3,289 engines in July 1967 to 2,515. Out of the original July 1967 figures for the RB211-06 estimate of 1,160 engines for the basic L-1011 aircraft, only 720 of these were for the basic L-1011 in the December plan.
32. **On January 28th, 1969,** the Rolls-Royce main board received a report from David Huddie, stating "The RB211 presents a formidable task for

the division and will take all our effort to produce a working engine on time". He indicated that the engine was nowhere near its performance requirements and there were still several serious mechanical problems. This followed the redesigned fan casing that had been modified, due to Lockheed wanting to alter the engine mounting, being found to be unsatisfactory and in need of further design.

33. **In March 1969**, a product profit plan for the entire RB211 project, including stretched versions, was produced. Figures for the production costs estimates in July 1968 (paragraph 26.) were presented in this March 1969 product profit plan and showed a more favourable position than that of the figures in December 1968 which showed a decrease in sales of the RB211-22 engines.
34. **On March 26th, 1969,** a meeting took place with National Union leaders regarding potential job losses.
35. **In May 1969**, the treasurer had been concerned that by the end of September bank borrowing would have reached £47 million compared with the profit plan of £35 million. As will be seen this sum was exceeded by September.
36. **On May 5th and 6th, 1969**, at the DED board meeting Tom Metcalfe, RB211 Programme Director, reported the latest estimates of the engine production costs. The projected, potential unit production costs had increased over the target costs, (against the cardinal rule of target costing) of some 12.6%, (although the design change orders (DCO's) were included in the production cost estimate, they had yet to be accepted by Lockheed) Future costs hazards were noted, mainly the titanium fan blade option, which would increase the cost by £24,000 per ship set (3 installed engines per set, 3 sets of power plant items, plus a set of pods–(two wing fitments and one tail fitment); the re designed fan case being dealt with by BED along with the weight reduction programme.
37. **On May 7th, 1969,** at the chief executive's meeting, Sir Denning Pearson, was advised of a 'small potential overspend' on the RB211 development costs projected for 1969 with no comment on the estimated increase in the production costs.
38. **By June 6th, 1969**, a revised RB211-22 total launch cost estimate was now £99.6 million (an increase of £10.3 million above that advised to HMG on 2nd September 1968, para 29).
39. **On June 30th, 1969**, Tom Metcalfe presented another report to the DED board (The Metcalfe Report). This isolated for the first time the profitability of the RB211-22 engine for the basic L-1011 aircraft. This

report was based on current costs rather than a target cost. The initial RB211-22 target cost was set at 8% lower than the production cost estimate based on the May 1968 design specification. The practice was for the target cost to be adjusted from time to time when necessary to take account of specification changes. The theory was that target costs would be practically achieved through cost reduction schemes. It showed the profitability of the RB211 project to be negative for the first time whilst the estimated costs associated with the programme had increased to such an extent that the previous profitability of the project was now showing a loss of £16.3 million compared with a planned profit of £46.9 million (an 'adverse variance' of £63.2 million).There had been a deteriorating position in expected sales of the RB211-22 from those presented in March 1969. The figures had not been approved by the heads of department and therefore open to challenge. Sir David Huddie believed some of the increases in administration, sales and service costs were 'untenable' and stringent targets for cost reductions had to be made. Rex Nicholson, concerned about the reduction in manpower against the effort required to produce the stated performance on the engine warned, there was . . . "no reserve of manpower to deal with any further changes to the engine specification . . . and no allowance for damages which might be claimed in the event of delay in delivery . . ." The minutes also record that the draft five-year forecast was 'clearly not acceptable' The figures were not formally agreed by the DED board. Neither the technical difficulties nor the financial problems were reported to the main board when they met on the 15th of July 1969 and it would later be said to the Inspectors, the 'Metcalfe Report' was not included in the DED papers to the main board members not on the DED board, which included Sir Denning Pearson. The five-year forecast was incomplete. The forecast for 1970-1974 had assumed some support from HMG for the stretched version of the RB211-22, the RB211-524, which was not confirmed. It showed the division's profits would drop considerably by £45.9 million to 1973, with an increase in R&D costs of £41.1million.

40. **On July 2nd, 1969**, the Chief Executives meeting was held. Sir Denning Pearson agreed that the draft DED five-year forecast should not be put to the main board since it had not been approved by the division's board and that plans for more acceptable results were being formulated. No mention was made of the Metcalfe report and the implications were not highlighted either at the next board meeting on the 15th of July 1969.

41. **On July 15th,1969**, the main board met. Sir David Huddie referred to the serious mechanical and performance problems with the RB211 but

stated 'in our judgment the task remains very difficult but achievable'. The DED forecast was confirmed as unacceptable although no figures were presented and that following a complete review of the division's activities, a revised forecast would be prepared by the end of August. The seriousness of the reported information set out in the Metcalfe report and the decline in profits compared with earlier forecasts was not conveyed. The DED board was aware of the serious financial and technical difficulties the RB211 project was experiencing, although the majority of the main board members were still not aware and would not be aware of the financial implications of the DED draft forecast until the October main board meeting.

42. **On August 18th, 1969,** At the RB211 programme meeting, John Bush, Engineering Director-RB211, requested that every engineering change should be approved by himself or the chief engineer with a review taking place daily. It was confirmed at this meeting that Lockheed and Rolls-Royce had reached the conclusion that progress on the Hyfil fan blade had reached the point for them to stop further development on the back-up titanium fan programme.

43. **On August 29th, 1969,** Launch-aid terms were finally agreed. HMG agreed to contribute retrospectively 70% of certain costs.

44. **On September 1st, 1969,** at the DED board meeting, Ernest Eltis, Director of Engineering, having described technical difficulties, delay and a 'tendency to overspend' on the engineering programme, asked that the RB211 engineers be allowed to concentrate on the development programme, having been involved in sales presentations at the expense of development. He also reported that the RB211-22 engineering costs were then estimated at £27.42 million compared to £23.95 million in the original profit plan for 1969. Tom Metcalfe RB211-programme director and previously Rolls-Royce production director, stated that modifications to achieve reliability and performance should take a higher priority over that of achieving weight reduction.

45. **On September 16th, 1969,** at the main board meeting, Sir David Huddie gave an account for DED stating, "Meeting our programme on the RB211 now represents a formidable task and I will be putting the programme on an emergency basis . . ." The specification still had not been finalised that would meet the contract commitments. The main board were made aware by this statement that all was not going well with the RB211 project. The DED revised five-year forecast would not be available until the end of September. Around the same time, Rolls-Royce had informed MinTech of its deteriorating financial position. MinTech then asked the Industrial Reorganisation Corporation (IRC)

to undertake an investigation into the company's financial position on its behalf. The IRC had been set up, amongst other things, to promote 'industrial efficiency and profitability'; it did not share companies' confidential information and therefore would report to MinTech but would only provide a summary of the company's financial position. Although the IRC would investigate thoroughly, MinTech would not have a detailed picture of the RB211 programme and the financial effect on the company. Back in May, the concerns over bank borrowing the treasurer had reported were confirmed with bank borrowing amounting to approximately £54 million. The Board agreed due to the financial position and bank borrowing it could not launch the RB211-50 without financial support from HMG.

46. **On October 1st, 1969**, the IRC began its investigation into Rolls-Royce's financial position.

47. **On October 6th, 1969**, a special meeting of the DED board was convened to discuss the five-year forecast which David Huddie reported as remaining unacceptable. The problem being the declining profitability in years-1972 & 1973. He decided that the forecast would be dealt with in two stages going forward, up to the end of 1972 and 1973-1974. Cost reductions in all areas continued to be countered by possibilities of increased costs in meeting the contract specification and negotiations with Lockheed on modifications, despite this it was decided a reduction in the workforce was needed. The board rejected the five-year forecast.

48. **On October 17th, 1969**, a provisional five-year forecast for 1970-1974 was sent to the IRC. It pointed to a far worse position, being half of what had previously been forecast in November 1968 through to 1972.

49. **On October 21st, 1969**, Sir Denning Pearson at a main board meeting advised that the new five-year forecast showed a serious decline in profitability. He stated that the IRC would concentrate on this in order to advise HMG on the extent of financial assistance it might require for launching the then RB211-50, included in the five-year forecast above.

50. **On October 24th, 1969**, Sir Denning Pearson held a special conference of senior management to discuss profitability. Cost saving schemes were introduced which included freezing senior salary increases, restrictions on Company cars and curtailing unnecessary office expenditure.

51. **On November 3rd, 1969**, at the DED board meeting, the board was told that the projected DED five-year forecast figures were worse than had been previously rejected by the divisional board on the 6th of October 1969. The company production director, Rex Nicholson stated it was proving very difficult to establish a factual manufacturing cost for the RB211 due to the number of recent design changes. In his view, quite

apart from cost, the whole programme was questionable due to the fact it was unlikely that engines of a suitable standard could be run or delivered to Lockheed by the stipulated dates, concerned that parts to the correct standard would not be available in time. (see paragraph 58). Ernest Eltis, director of engineering, reported no significant change in the serious position of the RB211 as had previously been reported on the 1st of September 1969.

52. **On November 12th, 1969**, the main board finance committee met to consider the five-year forecasts of all divisions, in particular the DED forecast. They had received a breakdown of the difference noted between DED's 1970-74 forecast and its 1969-73 forecast the 'adverse variance ' but wanted to know what action was to be taken to improve the situation. The committee did not approve any of the divisions five-year forecasts submitted. These were to be re-submitted with a plan of costs savings to be achieved through to 1974.

53. **On November 18th, 1969,** the main board met and was told the finance committee had not approved the five-year forecast. At the same meeting Sir David Huddie reported on DED, particularly the 'October variance'. He also believed the delivery of engines to specification to correspond with delivery of aircraft to passenger standard was still just achievable with 'all-out effort'. The forecast profits of DED had deteriorated compared with June 1969 and its cash needs had increased by £49.7 million compared with the 1969-73 forecast of November 1968.

54. **On November 27th, 1969**, a second meeting was held with national union leaders. Sir Denning Pearson outlined the changes in the business' prospects which had occurred since their first meeting in March. The possibility of plant closures would not be ruled out.

55. **On November 28th, 1969**, Tom Metcalfe had written to Sir David Huddie stating, ". . . I am convinced that unless we take some very special steps right away, you will not have an engine that works at all. We will be seriously late with engine deliveries and eventually will be the cause of delay to the aircraft programme and the aircraft deliveries to the airlines . . ."

56. **In December 1969**, the IRC representatives completed their investigation of the company and submitted a report to the IRC Board. Recommendations included the IRC should agree to make a convertible loan of £10 million available to Rolls-Royce; Rolls-Royce should agree with IRC's recommendations for strengthening the top management structure of the company; Rolls-Royce should sell OED; Rolls-Royce should provide the IRC with quarterly financial information on each division; Rolls-Royce should consider proposals to integrate SED and BED.

Most of the proposals were rejected at the 20th of January 1970, board meeting although an amended version would later be approved.

57. **Throughout 1969** at both the main board and DED board meetings, concerns had been expressed about the deteriorating financial position for both the current and future profits as projected in the five-year forecast, 1970-1974.

58. **On January 5th, 1970**, the RB211 programme director, Tom Metcalfe, reported to the DED board that immediately prior to Christmas an agreement to reschedule the delivery programme had been reached. Launch costs, including continuing development costs, had risen again to £107.8 million.

59. **On January 14th, 1970**, meetings, requested by Lockheed took place in Derby. Lockheed representatives were critical, particularly in relation to the programme management of the RB211 project. The Lockheed representatives cited nine criticisms amongst which was the 'failure' of Rolls-Royce to supply definitive cost data for design change negotiations.

    Later at a programme meeting, it was agreed that the inner casing of the combustion chamber would have to be redesigned, with the additional burden of incorporating the changes in the production parts. Requirements regarding engineering and production relating to engine cooling were also proving difficult. The Hyfil fan blade problems of strengthening the root blade and the effects of bird strike were also discussed.

60. **On January 23rd, 1970**, the titanium fan blade back-up programme was re-established from its cancellation on August 18th, 1969, due to continuing problems with Hyfil and the potential of delaying the first flight of the L.1011.

61. **On February 9th, 1970**, Following on from the criticisms by the Lockheed representatives in January, Rex Nicholson was appointed as RB211 Programme Director and main board member, replacing Tom Metcalfe. He had two priorities ;1) to update launch costs, 2) to examine the estimated production costs. He was quoted as saying the RB211-22 "was a different animal from the one quoted on and I knew there was trouble there".

62. **In March 1970**, a new RB211-22 launch costs estimate was issued following an 'exhaustive reappraisal' of the situation. This showed that the estimated launch costs had increased from £89.3 million as advised to HMG in September 1968 to £134.7 million. This represented an increase of £26.9 million from January 1970. What had led to a review

of the engine delivery dates (see paragraph 58) also led to a revision of the development programme.

Difficulties continued to be experienced with the composite Hyfil fan blades. On an open-air testbed run, the Hyfil blades suffered a catastrophic fan failure. It was no longer possible to rely on the Hyfil fan blades for production and by May it was decided to abandon them, leaving only the titanium fan blade substitute which had been developed as a backup with a considerably greater cost and increase in weight resulting in other components also having to be strengthened.

63. **On March 2nd, 1970**, Ernest Eltis presented a report to the DED board stating that the "grave situation with the RB211 project had not improved". Due to the several technical problems outlined by him, 'no engine of a standard fit was produced for significant endurance testing'.

64. **On March 11th, 1970**, the finance committee of the main board met, with a representative of Lazards present. The finance board recommended accepting the offer from the IRC of a convertible loan of £10 million.

65. **On March 17th, 1970**, the IRC offer was put to the main board. However, the letter of acceptance was left to 'lie on the table' due to the IRC not being willing to confirm their offer until they had seen the company's five-year plan; HMG wouldn't decide on the RB211-50 until it had also seen the company's revised five-year plan and the Lockheed position clarified. Sir David Huddie still believed there was a good chance of meeting the programme, despite the difficult development situation and the programme having used up all the contingency time built into it.

66. **On April 17th, 1970**, Rex Nicholson presented the revised launch cost estimate to the main board (see paragraph 62) and having called for the revised production cost estimates for an RB211-22, installed engine (still with the Hyfil fan blades), these were also presented. These showed the costs had reached £205,380, based on an estimate issued on the 10th of April and were some £40,000 in excess of the target costs. Some £32,000 of this £40,000 excess was down to design changes. From the production cost estimate of the design at the time of the contract, (£153,655) this estimate was £51,725 more. The production cost would also increase by a further £12,000 if titanium fan blades were required. A representative from Lazards was present. On reporting the serious cash position disclosed by the revised five-year forecast, Lord Poole, Lazards' chairman advised Sir Denning Pearson in a lengthy letter, dated the 22nd of April, of the urgent action he believed was required, recommending a sub-committee and appointing a Mr. Hatch to represent Lazards as he saw it, 'to be involved in the very difficult

policy decisions which would need to be taken'. Concern was expressed as to how the major banks would react in June when Rolls-Royce was due to renew its financial facilities with them.

67. **On April 21st, 1970,** new RB211-22 production costs became available. These showed that the estimated production cost of an installed engine still with Hyfil fan blades had increased from the £205,380 notified on April 17th to the main board, to £217,338. This estimate was used in a new project profit plan which became available on the 24th of April 1970. The launch costs of the basic RB211 including continuing development, were now estimated at £151.8 million (£17.1 million higher than the March estimate).The figures in this plan included an 'uprated' version (RB211-50) of the engine and assumed sales of 385 engines as well as Industrial Marine division applications with assumed sales of 150 engines. These figures were fully escalated over the lifetime of the project. The figures concluded in the Metcalfe Report were for the RB211 alone and were not escalated. The new figures showed an increase in profitability over the forecast life of the project,(some 15-20 years) however, escalation figures aside, along with increased selling prices, the profitability was probably down to the inclusion of the additional estimated sales, offsetting any increase in costs. However, Mr. Trevan- Hawke, the company financial controller and a main board director together with the auditors were becoming concerned about the recoverability of the whole RB211 continuing development costs, which were being carried forward as an asset. The IRC were also aware of the deteriorating financial position and escalating launch costs of the RB211.

68. **On April 22nd, 1970,** Mr. Trevan-Hawke sent the five-year forecast (1970-1974) of the 17th of April to the IRC along with two schedules, showing original and revised figures which evaluated a plan concerning the Trent engine and indicating a new project. This would be reported by Sir Denning to the main board on the 5th of May 1970, that HMG and the IRC had been informed that there was no prospect of the company participating in the financing of any new major project until after 1972.

69. **On April 24th, 1970,** during a meeting with Mr. Trevan-Hawke the IRC suggested Rolls-Royce should admit that some of the development expenditure on the RB211 was irrecoverable and the company should write these off as a loss instead of carrying forward on the accounts. This was pressed home by Mr. Trevan-Hawke at a meeting on the 28th of April with Sir Denning Pearson, Sir David Huddie, Messrs. Hinkley and Flemming and the company's auditors, Cooper Bros.& Co, requiring

a provision of £20 million to be made in the 1969 accounts against the non-recovery of the RB211-22 R&D expenditure in an attempt to recognise the various matters which might affect the project during the forecasted 15-20 years of its lifetime.

70. **In May 1970**, Lockheed was experiencing weight difficulties with the airframe and sought to remedy the problem by asking Rolls-Royce for an increase in engine thrust from 40,600lb. to 42,000 lb. to counter this by March 1972, a year earlier than Rolls-Royce had planned to do so by September 1973.

71. **In May 1970,** Ian Morrow was approached by an IRC representative to consider joining the Rolls-Royce main board in order to strengthen the financial management of the company following on from the amended loan conditions agreed with the IRC. As a partner in the accountancy firm of Robson, Morrow & Co, and a respected and distinguished management accountant, he had also been involved in work for the Ministry of Supply. Prior to any decision, he requested the DED 1970-1974 forecast for analysis, which was consolidated with the group forecast and supplied to the IRC by Mr. Trevan-Hawke on April 22nd, 1970. Following this request DED analysed this for the first time so as to show the profitability of the RB211-22 alone. Whilst from a business forecast it showed profitability at the gross margin level for 1970-74 of £39.4 million, after loss before amortisation and charging amortisation of R&D costs, HMG levies and all selling and administrative costs, it showed a loss of £43.4 million.

72. **On May 4th, 1970,** at the DED board meeting Ernest Eltis reported that the RB211 programme situation remained very serious. The total engine running time was behind schedule, the main cause being the unsatisfactory state of the combustion chamber. Other known technical problems were high pressure turbine blade fatigue, excessive turbine seal excursions, fan blade vibration and high-pressure nozzle guide vane deterioration. He also confirmed the reinstatement of the titanium fan blade programme. Rex Nicholson reported the output position for 1970 looked poor and one of the principal causes was the high rate of design change for the RB211 affecting non-RB211 component production. Mr. J. Wood, commercial director, reported a general lack of market activity and little prospect of enquiries for sales, materialising in the short term, for the standard L-1011 aircraft.

73. **On May 12th, 1970**, the DED Commercial Director, Mr. J. Wood, pointed out to Rex Nicholson that whilst the increase in thrust would help the airframe weight it would only moderately solve the engine overweight. Mr. Wood also thought several concessions should be

required from Lockheed in exchange for this including an increased price of installed engines to committed customers and the removal of Lockheed's right to withhold $45,000 per engine in the event of engine performance deficiencies.

74. **On May 14th, 1970,** Rolls-Royce prepared a Memorandum of Understanding regarding the provision for increased engine thrust of 42,000lb. from the RB211-22 at a day temperature of 84^0 by March 1972. Commercial negotiations were not finalised by the time of the receivership. In the event, Lockheed never signed this document and Rolls-Royce were not committed to producing the increased thrust, although the engineers continued to work toward developing this capability, assuming the commercial negotiations would be satisfactorily concluded.

75. **On May 19th, 1970,** the board approved the amended terms and conditions from the IRC accepting a £10 million loan (the first loan) over 7 years, with interest. A second loan of £10 million would be considered, subject to the company not defaulting on the terms of the first loan which involved a plan of reduced short-term borrowings by the first of April 1971 of not less than £15 million. Furthermore, it would need to convince the IRC it actually needed the second loan taking into account other sources of finance available for the purpose of continuing R&D. Any application for the second loan needed to be received in writing by 31 December, 1970.

76. **On May 28th, 1970,** Rex Nicholson wrote to Sir David Huddie warning the programme was at that moment in 'grave jeopardy' due to a shortage of design effort to deal with, particularly, the requirements of weight reduction, gas generator fairings and thrust reversers.

77. **By mid-June 1970,** Sir David Huddie had collapsed with exhaustion and his medical condition meant that he never returned to work at Rolls-Royce. This was seen as a major blow to the management and running of the company in 1970.

    Fred Hinkley was appointed to act as the Managing Director of DED in the absence of David Huddie. The RB211 was by now 6-9 months behind schedule.

78. **On June 18th, 1970,** there was a change of government, and the new Conservative Government was advised about the difficulties facing Rolls-Royce. Sir Denning Pearson would write to the new Minister of Technology in July seeking to confirm the government's continued support for its R&D programmes.

79. **On June 22nd, 1970,** the company's annual accounts were circulated to shareholders and showed the special provision of £20 million had

been made to ensure the risk of non-recovery of the total research and development costs for the RB211.

80. **On July 6th, 1970**, at the DED board meeting, Tom Metcalfe's view, supported by Ernest Eltis and John Bush was that there was only a slight chance that the company would be able to supply engines to Lockheed on time with the thrust required. Ernest Eltis reported that . . . "the most economical method of containing a titanium fan blade in the event of a fan blade failure had not been established" and there had been very little improvement in the turbine area which was mainly responsible for the engine's poor performance. Rex Nicholson reported that the RB211-22 launch costs had been reassessed at £148.5 million (excluding continued development) and could be more; £20 million higher than indicated in a product profit plan of 24th April at a comparable stage of development. Fred Hinkley believed there wouldn't be any great improvement until 'a stabilised design was achieved'.

81. **On July 14th, 1970**, a revised RB211 launch cost estimate became available. The estimate of £143 million was close to the £148.5 million, excluding continuous development costs in both estimates, forecast by Rex Nicholson at the above DED meeting on the 6th of July. When these were added, the estimate totalled £169.2million. When projected through to the Project Profit Plan (PPP), the project was estimated to return a loss of £61.6 million and up to £98.9 million if further costs, mainly in warranties and guarantees, were added. An appraisal of what costs for sales of spare parts to set against this loss would not be available until autumn of 1970 due to the engine design being subject to so much change. The plan projected sales of 1,217 engines for the RB211-22 for 338 aircraft and 150 engine kits for the industrial and marine market.

82. **On July 15th, 1970**, at the mainboard finance committee meeting, Mr.Trevan-Hawke presented the above profit plan and produced broad calculations based on that plan showing the extent of the estimated loss arising on RB211 business in the years up to and including 1974. It showed a Gross margin of £19.2 million and a loss before amortisation of R&D and development costs of £63.1 million, with amortisation of R&D and development costs of £11.4 million, totalling £74.5 million. Losses before amortisation of R&D of some £54.6 million were estimated to occur mainly in the years 1971-73. This was £31.1 million worse than included in the 1970-74 five-year forecast presented to the IRC on the 22nd of April 1970. The report assumed that the 2nd tranche of the IRC loan of £10 million and a reduction in the company's overdraft by £15 million would be achieved. However, the

company would require bank and acceptance facilities in excess of the £70 million in 1970 and more so in years after.

Despite other areas of the business being profitable, the RB211-22 programme was not proving to be so. The directors also considered whether to disengage from the L-1011 programme and whether anything should be said about the estimated figures produced by Mr. Trevan-Hawke at the forthcoming AGM. Mr. Ian Morrow, who had attended, thought that something should be said. Mr. Hinkley stated that plans were in hand to reduce the production cost of the RB211-22 engine by £30-£35,000 per engine.

83. **On July 21st, 1970,** the main board at their meeting considered the above-mentioned report which showed a need for an estimated £30 million, which might be raised internally through leasing of plant and disposing of certain assets along with a reduction in manpower and stocks. Fred Morley was appointed to the DED as director of design to strengthen the design function of the engineering department. In line with the IRC conditions, Ian Morrow was appointed as Deputy Chairman to the main board. Lord Beeching was also appointed to the main board on August 1st, to represent the interests of the IRC.

Sir Denning Pearson addressed the shareholders at the AGM on the same day, with a statement agreed by the board. On the basis of the figures made available to the board, he was advised that he needed to qualify certain comments he had made in his previously published statement to shareholders as to the adequacy of the company's working capital. Furthermore, with regard to profitability, he should give 'very pessimistic comments' about the results for 1970-1972 so as to avoid embarrassment at subsequent shareholder meetings. Lord Poole of Lazards, had told Sir Denning if he was not prepared to say the company could carry on financially, the board would have no alternative but to appoint a receiver. However, as these were provisional figures, Lord Poole thought it would not be in the interests of the shareholders or creditors to place the company in liquidation that weekend but believed with the £20 million loan from the IRC (of which only £10 million had been agreed at that point) and were there a reasonable chance the launch aid agreement could be renegotiated so that HMG would contribute to the overrun of development costs, the company could absorb the production losses. In those circumstances he believed it would be possible to say that the company had the financial ability to carry on trading. There was no mention by Sir Denning to the shareholders, of the substantial loss the RB211 programme was estimated to incur as was outlined in the most recent provisional figures or the huge cash shortages that were indicated. Although MinTech were aware of the technical difficulties of the RB211-22, they were not aware of the liquidity problems the company were facing.

84. **On August 10th, 1970**, a new RB211-22 production cost estimate was available. The estimated production cost of an installed engine had increased to £237,452 (an increase of £20,114 since April 21st -paragraph 67), showing an increase of £83,797 overall from the original estimate prepared in July 1968 (£153,655.00).

85. **September 17th, 1970** was the first main board meeting that Ian Morrow and Lord Beeching for the IRC attended. The meeting had been called to discuss the state of the RB211 project and financial implications. The company's bankers had expressed concern that the company was using the overdraft facility as permanent capital.

86. **On October 12th, 1970**, the new five-year forecast was issued, accompanied by a statement and copies sent to MinTech and the banks. Slight amendments were made by the company and incorporated in a forecast dated October 14th, 1970. This forecast showed that after including a margin of £20 million for contingencies, a peak cash requirement of £119.5 million would arise at the end of 1972. This was £49.5 million more than the company's agreed overdraft and acceptance facilities of £70 million. However, the statement allowed for a further £5.5 million by way of an additional contingency margin taking the total cash requirement to £55 million.

87. **On October 13th, 1970**, at the main board meeting, it was stated that as a result of a major revision of the RB211-22 programme in the previous few months, a complete reappraisal of the launch costs was in progress.

88. **On October 29th, 1970**, the main board met again and were appraised of the two meetings, the first with the Ministry of Aviation Supply (which had taken over MinTech) on the 27th of October and the second with the Governor of the Bank of England, Leslie O'Brien, on the 28th of October; both had been inconclusive but both would continue to have discussions. Sir Denning had been informed by the Governor that if financial help was to be forthcoming from HMG or the private sector there would need to be changes in management and required that Lord Cole be appointed as chairman. Sir Denning recommended Lord Cole to the main board.

89. **On October 30th, 1970**, Lord Cole went to see the Prime Minister, Edward Heath, stating that he was not prepared to undertake the Chairmanship of Rolls-Royce unless he was assured that the financial aid required would be supplied. Lord Cole became Chairman on November 1st, 1970.

90. **On November 3rd, 1970,** a letter outlining the discussions and proposal put to Lord Cole by Prime Minister Heath was received stating, in view of the damage to the National interest which would result from Rolls-Royce defaulting on its contract with Lockheed, he would regard it as reasonable for Rolls-Royce to ask the Government to meet part of the overrun on development costs for the engine. Given the previous Labour Government had committed to providing launch aid of 70% of the original estimated cost of £60 million relating to R&D, the Government would be prepared to play a reasonable part in dealing with the overrun, however it would not be prepared to meet it all and neither could it accept an open-ended commitment. It went on to say that the Government would be prepared to adopt a reasonable approach towards contributing to any further escalation of these costs, provided the reasons and need for the escalation were fully agreed with the government. Equally, the payment would be adjusted downwards in the event less cash were required. This would remove the need for the second tranche of £10 million of an IRC loan. An independent firm of accountants would be engaged to advise the Prime minister that the proposed Government contribution was justifiably related to the cost of the R&D attributable to the RB211-22. Rolls-Royce were expected to implement changes to the management structure and to give the accountants open access to the books.
91. **On November 9th, 1970,** Sir Henry Benson, senior partner at Cooper Bros & Co. Accountants, had been approached by HMG to provide a report by this date. Sir Henry Benson submitted a report but pointed out there had been insufficient time to probe or verify the Company's estimates. HMG made it a condition of any aid that an independent examination of Rolls-Royce take place. Although the immediate problem of the proposed additional launch aid was resolved, there were a number of long-term problems which required examination.
92. **On November 11th, 1970,** HMG publicly announced its willingness to provide additional launch aid to 70%, of £60 million, to a maximum of £42 million subject amongst other things, to independent financial verification of the R&D figures by independent accountants. The sum of £60 million, representing the approximate difference in the increased estimate of RB211 launch costs of £137.5 million from the original £74.9 million (£62.6m). The two clearing Banks, Lloyds & Midland had agreed to provide further funding of £10 million each (which would be added to a reduced overdraft facility of £20 million each on acceptance of the terms) with the Bank of England agreeing to provide £8 million. Rolls-Royce already had a credit facility of £20 million. The total would make £88 million available in short-term facilities.

The deficit in launch aid would be made up by the two clearing banks providing £5 million each out of the funding along with the £8 million from the Bank of England.

93. **On November 16th, 1970,** the DED board met and although they discussed proposed economies of £5 million per annum in wages and salaries by the end of 1971 as a target and redundancies, no mention was made of the RB211-22 recorded in the minutes.

On the same day the first flight of the L-1011 took place as planned with the 'Batch 1' development engines which had achieved the required thrust for that stage. The next day Mr. J. Bush would inform Lockheed that the 'Batch 3' engines had thrust deficiencies and additional design changes were being made for engines 10010 & 10011.

94. **During the autumn of 1970,** questions were being asked in Parliament as to whether the total of £89 million (the original £47 million and the further £42 million) would be adequate and whether in fact HMG's commitment would reach £150 million. Mr. Trevan-Hawke would write to the divisional financial controllers, the last paragraph of which indicated the £42 million additional launch aid would not be drawn on until the report by Cooper Bros & Co. had completed the examination.

95. **On November 24th, 1970,** Cooper Bros. & Co. was formally asked by HMG to investigate the finances of Rolls-Royce Ltd., to examine and report to the Minister of Aviation and Supply on the profit and cash forecasts of the Rolls-Royce group of companies for the years 1970 to 1974, however in the course of their investigations, the 1971-1975 forecast was being prepared based on then current information and subsequently Cooper Bros.& Co. decided to review those forecasts which would not be available until the middle of January 1971 when they would be submitted to the main board for approval on 26th January 1971 and so the completion of their examination would be delayed, although their work continued throughout the following months.

96. **On November 30th, 1970,** the Director and Chief General Manager of the Midland Bank Limited (eventually taken over by HSBC) had written to Lord Cole outlining the terms on which the additional bank facilities of £28 million (a loan of £10 million from the two clearing banks and £8 million from the Bank of England) would be made available to it. The letter would state that when formal agreement to the arrangements had been received, Lloyds Bank Limited and Midland Bank, who were providing the overdraft facility would be advised and the overdraft for each would immediately reduce to £20 million from £25 million, on the usual banking terms and conditions. Whilst indicating the additional facilities would simultaneously become available to Rolls-Royce on

the acceptance of the terms, as far as the banks were concerned, this was dependent upon the results of the report by Cooper Bros. & Co's examination of Rolls-Royce's financial affairs, and thereby HMG providing the £42 million additional launch aid. However, as Lord Cole responded on the 7th of December confirming acceptance of the terms, the company was now, as far as the banks were concerned, immediately operating at reduced bank overdraft facilities totalling £40 million (from £50 million). No mention of this reduction would be made at the main board finance committee on the 11th December 1970. Lord Cole would later state that he believed the new additional facilities would be available from the point of acceptance of the terms.

97. **On November 30th, 1970,** the DED board met. Discussions took place on the draft five-year forecast (this didn't include the RB211-61 which HMG would shortly announce it would not support) and accepted that further amendments were required. A number of reports were presented to the board in the 'Red folder'.

Mike Eltis as director of engineering, reported amongst other matters, that the area of most concern was the high temperature technology and further modifications were needed to improve capability to the necessary level. The failure to obtain the required thrust linked to this was thought to be an inability to run the engines at higher temperatures. Additionally, a serious life limitation had been exposed on one of the HP compressor discs. Eric Scarfe, divisional production director, in his report, expressed concern at the number of modifications being called for in engineering with the inherent effect on the production department. He warned of the 'swingeing effects in 1971 on all aspects of the business and the effect would be measured in millions over and above that which had been planned for'. Rex Nicholson, programme director, confirmed the L-1011 had flown successfully on the 16th of November 1970. Although the development programme was gaining momentum, the engine specification still had not been finalised. He outlined four significant problems as a consequence of the low thrust available on the first flight (batch1) causing a reassessment of the programme of engines to be delivered subsequently for the flight certification programme increasing from 30 to 44 engines. This would affect the production output rate in the first half of 1972 because delivery rates would need increasing. A higher rate of spare parts requiring additional capital investment was forecast and lastly, support from the bench development and flight certification programmes would be needed to make the changes necessary to achieve an engine which would pass aircraft certification.

Although a re-estimate of the launch costs of the engine in the light of the changes he'd outlined had not been completed, neither the potential increase in launch costs nor the effect on the production department of the necessary changes referred to, were allowed for in the draft five-year plan. The draft had also assumed that from the £37.8 million additional launching aid (£42 million minus £4.2 million refund of investment grants), £19.7 million would be received in 1971 with the balance of £18.1 million, in 1972, but this was not apparent from the forecast which would be examined by Cooper Bros & Co. The concerns outlined by Rex Nicholson were requested by Mr. Conway to be drawn to the attention of the main board.

98. **On December 4th, 1970,** a production cost estimate was produced which showed the cost of the RB211-22 engine had increased by £2,500 per installed engine from the August cost estimate, prepared for the 17th of September brochure presented to the Ministry of Aviation Supply.

99. **On December 9th, 1970,** Cooper Bros & Co., were supplied with a copy of the RB211-22 product profit plan. It was described as 'not reflecting of the situation to date but rather the situation upon which the provision of the £35 million in the company's interim statement was based'. Assumptions had been made as to forecasting sales both within the aircraft industry and Industrial Marine applications. All estimates were overly ambitious as it transpired.

100. **On December 17th, 1970,** at the main board meeting in London, Mr. Geoff Fawn's appointment as managing director of DED was announced to the directors. Lord Cole and Messrs. Conway, Nicholson and Morrow all reported on the meetings with Lockheed which had taken place in the US between the 12th and 15th December 1970, which they considered satisfactory in strengthening relations between the two companies. They reported that Lockheed was concerned regarding Rolls-Royce's ability to meet the engine requirements but 'believed no effort would be spared' and that they would be successful. However, Lockheed did not appreciate that there would be no financial rescue by HMG, should Rolls-Royce have further financial difficulties. There were no recorded comments in the minutes on the treasurer's report, contained in the 'red folder', which stated there had been no official advice of a change in borrowing limits. This was despite previous discussions proposing to amend the limits with the bankers and therefore implied, they remained unaltered at £50 million from the clearing banks and £20 million from the acceptance houses. No reference was made to Lord Cole's letter to the Midland Bank Limited earlier in December.

Rex Nicholson in his report to the board stated that when discussions had taken place with Ministry of Aviation Supply which led to the £42 million additional launch aid being agreed, the Ministry had been advised that the estimates did not reflect the costs of the revised programme not at that point discussed wuth Lockheed. Although, it's unclear as to whether the Ministry were aware of any re appraisal of the development programme or moreover the re estimation of the launch costs discussed at the 30th of November DED board meeting, referred to by Rex Nicholson; Mr. Morrow was recorded in minutes of a meeting held on the 2nd of November with the Ministry, confirming as still valid, the launch cost estimate when he'd met ministry officials in September 1970. With regard to manpower, reductions were confirmed at approximately 4,000 since the 1st of November 1970 and there were plans to reduce further by at least 3,200 by the end of 1971.

101. **On the January 1st, 1971**, Lord Cole met with Sir Henry Benson of Cooper Bros.& Co. who had been examining the group forecasts for 1970-1974 and the DED 1971-1975 draft forecast of 30th November 1970 and had been comparing the past performance of the company and DED with the relevant forecasts. They had concluded that the forecasts were over optimistic and reflected management targets rather than likely performance. Sir Henry was concerned that the forecast was too 'sanguine' and that the board would likely approve the forecast profits for 1971-1975 when it was presented on the 26th of January 1971, before his firm had completed its review and he may then disagree with the forecast figures. Lord Cole concurred and believed they should be reduced by a half or two thirds. They agreed to discuss the figures and board approval would not be sought until they had had a further conversation. Sir Henry Benson had around this time formed the opinion that the £60 million additional finance negotiated in November was unlikely to be sufficient and thought at least a further £20 million might be required and had taken into account the approximate amount of £13 million cash requirement which, Mr. Morrow had earlier advised Lord Cole, had been understated in the 14th October 1970 cash requirement forecast.

102. **On January 11th, 1971**, a revised draft DED 1971-1975 forecast became available, the implications of which were discussed the next day by Messrs. Conway, Trevan-Hawke and Morrow. With no contingency margin included, it showed an increased demand on Head Office of £62.6 million by the end of 1973 with peak demand of £47.8 million compared to that previously forecast when the £60 million of launch aid plus banking facilities were being negotiated. Therefore, the

contingency margin of £23.3 million in the £60 million was more than absorbed. This amount was required to finance further R&D on the RB211 of £20.4 million and cover the reduction in profits of £42.2 million.

Two separate reports by Rex Nicholson outlined that there were still difficult problems with the engine development as to weight, integrity, performance and component life. Whilst it had been previously considered twice before, it was now clear an extension to the time for completion of the contract would be necessary as it became apparent that Rolls-Royce would have difficulty in meeting the Lockheed programme.

103. **January 18th, 1971**, as would transpire, was the last date that some of the DED board met as had been requested by Geoff Fawn on the 8th of January. It has been timetabled to talk about the RB211-22 problems and engine weight particularly. Stanley Hooker was to be brought back from retirement and appointed group technical director. The RB211-22 product profit plan became available. Mr. Conway commented on the disastrous implications of the 1971-1975 five-year forecast. However, with the exception of certain matters the forecast was accepted. Although the RB211-22 product profit plan was available on this day, this was not discussed. Amongst other matters, this plan showed the effect of the levies that would be paid against the additional aid of £42 million, which would reduce the extent of profitability over the life of the project. It would also show that the estimated launch cost of the project had increased by £32.4 million to £202.7 million since the figures contained in the brochure dated 17th September 1970, presented to the Ministry of Aviation Supply. The plan differed to that presented to Cooper Bros & Co. in a number of factors including an escalation of income and expenses for inflation during a further year, an assumption that the commercial life of the engine would be five years longer and therefore included estimated sales of spares but excluded marine and industrial sales, which had been included in the plan presented to Cooper Bros & Co.

Mr. Eltis stated the weight of the engine was not realistic in relation to the commitment to thrust and the most serious problems remained in the areas of High temperature technology, particularly the first stage nozzle guide vane (NGV), first stage turbine blade, control of turbine seals at high temperatures, the combustion chamber, and high-pressure turbine blade vibration. A critical period in case of any further need for changes or re-design of parts was thought to be nine months, according to Rex Nicholson. The problems with the Batch 3 engines remained unresolved. Following this meeting, an assessment by Geoff Fawn thought the RB211 programme would have to be delayed by at least six months. A meeting of executive directors would later decide that the delay would be for six to twelve months.

104. **On January 19th, 1971**, the executive committee met and discussed the 1971 profit plans except those of DED. Lord Cole, with Messrs. Morrow, Conway and Straight met with the Minister of Aviation Supply with discussions centering on international cooperation. Given the main board finance committee had not been advised of the up-to-date position, no mention was made regarding the need to defer the programme. Later that day the group 1971-1975 five-year forecast became available. This forecast did not include contingency margins. The consolidated forecast showed a peak financial requirement of £55.6 million more than had been forecast on the 14th of October and £7.8 million more than reflected in the DED forecast that had been discussed by Mr. Morrow with Messrs. Trevan-Hawke and Conway only seven days earlier on the 12th of January 1971. A peak requirement would arise in 1971, of £121.9 million after receiving the available additional launch aid. The £88 million in proposed bank and acceptance facilities would be insufficient by £33.9 million. The delay to the programme would be conveyed by Lord Cole by Mr. Morrow that evening.

105. **On January 20th, 1971**, the Finance committee met. Mr. Morrow reported a 'substantial deterioration in the future financial position of the company', but no up-to-date figures were given. Sir Maurice Laing, a non-executive director, told Lord Cole that the Governor of the Bank of England should be informed and that he was meeting him that evening. Lord Cole would inform Sir Henry Benson of Coopers of the position also that evening. There seemed only two courses of action open, a) to stop the RB211 programme altogether or b) to negotiate for further time in which to complete the RB211 programme. The problems arising from either were, if the RB211 programme were stopped, the basis for the recently agreed additional launch aid would disappear and if the project were delayed there would be financial implications on late delivery. Either way the position was extremely serious.

106. **On January 21st, 1971,** Sir Henry Benson of Coopers assisted Rolls-Royce in producing a memorandum to present to the Ministry of Aviation Supply advising them of the serious situation and setting out tentative figures and factors of stopping or postponement of the programme and either a moratorium for creditors or liquidation of the company were set out. Some of the figures sent to head office by DED were hastily calculated as to the effect on DED's cash requirements of extending the

programme deadlines. The Memorandum referenced the need to exceed the 'existing facilities of £50 million of bank overdrafts and £20 million of acceptance credits', this despite Lord Cole's acceptance letter to the Director and Chief General Manager of Midland Bank Ltd. of the 7th of December 1970, which reduced the same by £10 million overall. The Ministry became aware of the extent of Rolls-Royce's problems, both financial and technical, the next day.

107. **On January 22nd,1971,** Mr. Trevan-Hawke heard of the November 1970 reduction in overdraft facilities from £50 million to £40 million, four days before the main board meeting was to be held. On the same day Lord Cole, Mr. Morrow and Sir Henry Benson informed the Permanent Secretary of the Ministry of Aviation Supply, Sir Ronald Melville of the technical difficulties of the programme, a possible delay of six months in delivering the first engines and the financial input that was needed to overcome these difficulties. In addition, he was told that the board would probably terminate the RB211 programme. No further pitch was made for additional launch costs which were unlikely to save the company from the enormous, estimated operating losses and potential liability in damages from Lockheed and the airlines, which were considered largely unquantifiable. On leaving the ministry, they went on to see the Governor of the Bank of England, then on to Lord Poole of Lazards, who represented the accepting houses.[13] The Treasury was informed of Rolls-Royce's predicament.

Further discussions took place later that evening between Lord Cole, Sir Henry Benson and Mr. P.G. Benham of Linklaters & Paines, Solicitors, acting for Rolls-Royce. Mr. Benham concluded if the additional cash requirements were not agreed or Lockheed did not agree to the six-month delay as set out in the memorandum, it would certainly end in receivership, as there was a serious risk that Rolls-Royce Ltd. would be in breach under section 332 of the Companies Act, 1948, by trading insolvently.

108. **On January 23rd, 1971,** the Chancellor of the Exchequer informed the Prime Minister of Rolls-Royce's position.

109. **On January 25th, 1971,** The Parliamentary Secretary for the Minister of Aviation Supply, who was out of the country at the time, held a meeting with officials, representatives of the Treasury and Sir Henry Benson. Following this he reported to the Prime Minister later that day

13 A financial institution that 'accepts' or guarantees a bill of exchange, thereby providing a guarantee about the creditworthiness of the drawer of the bill provided, subject to a commission.

that the future of the RB211 and Rolls-Royce itself was at risk. Further major financial intervention by HMG may be required to protect defence projects. The board was meeting the next day and their view was awaited. A draft RB211 product profit plan was produced on the same basis as the previous plan but incorporating the effect of a six-month delay and assumed due to the delay sales of 135 engines would be lost-but no mention of any potential claims from either Lockheed or the airlines.

110. **On January 26th, 1971,** Rolls-Royce main board meeting concluded that the £60 million additional finance negotiated in November 1970, would no longer be sufficient and £100 million would now be required. Further discussions with HMG were to continue over the following few days. The Board heard presentations from Mike Eltis supported by Stanley Hooker of the technical difficulties encountered by the engineers and the inherent delay if the difficulties were to be resolved. The Board further concluded that although the technical problems were not insurmountable, the escalating costs were unlikely to be met by either current resources, or outside assistance. Ian Morrow reported the reduction in overdraft facilities provided by the Midland and Lloyds Banks to the Board. This coupled with the implications for delay and possible claims by Lockheed, meant the Board, subject to negotiations with Dan Haughton and HMG on the contract, would be left with no alternative but to stop the project.

This was the first time some board members had been made aware of the real gravity of the situation the company was in and receivership as a possibility. Many were reported as being stunned. Mr. Trevan-Hawke, finance director, who had only heard of the November 1970 reduction in overdraft facilities some four days before the board meeting, had in his financial report to the board referred to current borrowing limits of £70 million (which included £20 million financial acceptances (a short-term credit instrument requiring payment when due), as continuing to be available until the additional launch aid was confirmed. However, the company faced the immediate problem of not being able to pay creditors as debts were becoming due. The overdraft was close to the £40 million of the reduced overdraft facilities. There was little time left for the board to consider any alternative course of action. There had been no previous indication to the Board of the need to keep within the overdraft limit and it had not been brought to the attention of Mr. Trevan-Hawke.

111. **On the afternoon of January 26th, 1971**, a series of discussions took place at the offices of the Permanent Secretary of the Ministry of Aviation

Supply. The Board's decision from its meeting was reported. It was hoped that Dan Haughton could be persuaded to either terminate the contract or agree to modify it in Rolls-Royce's favour. If Dan Haughton was willing, it was still hoped that a viable solution could be arrived at. In order to negotiate Rolls-Royce needed to have some indication of the stance HMG would adopt

112. **On January 27th, 1971**, Ministry representatives were sent to Derby to assess the technical state of the RB211. Completed by the next day, it concluded all the technical problems were capable of resolution over time but confirmed that an additional six months extension on contracted delivery dates was necessary to resolve. At a meeting between Sir Henry Benson, Lord Cole and representatives of the Midland and Lloyds Banks, the banks refused for the old limit of £50 million to be restored. However, the banks did agree to open a special wages account.

113. **On January 29th**, 1971, Lord Cole, Ian Morrow and Sir Henry Benson saw the Governor of the Bank of England to update him on the situation. Were the RB211 to continue, a further £150 million would be required which would include £45-50 million estimated as a provision for claims by Lockheed and the airlines. The same day, having received the report on the technical excellence of the RB211 engine, HMG views became known as would later be recorded in Cmnd 4860 (para's 27-29); Rolls Royce's liabilities were 'large, hurriedly estimated, in part unquantified and unquantifiable'. The paper went on to say that it would be an irresponsible use of public funds to take over the company and absorb the debt and obligations or support Rolls-Royce with funds it could not repay. The two choices it was left with was to take no further action and let the receivership and ultimately the liquidation of the company to take its course or to enter into negotiations with the receiver for those assets vital to the national interest, the decision was to be ready to take the latter.

114. **On January 31st, 1971**, (Sunday) Lord Cole met with the Minister of Aviation and Supply and the Permanent Secretary to discuss the probable timetable of events. It was still hoped that a favourable response from Lockheed might be received that would save the company from receivership. Lord Cole stated that strictly the company was already insolvent. Legal Counsel's opinion was that trading should stop. Special wages accounts were opened, and instructions given not to pay creditors. The Minister expressed the view that it did not give much time to hold the situation even if Lockheed responded favourably. Rolls-Royce reported a peak cash requirement of almost £122 million in 1971 so the company's proposed lending facilities of £88 million would

be insufficient by nearly £34 million. Lord Cole emphasised the major problem at that time was the company's immediate shortage of cash, although acknowledging potential claims from Lockheed. Lord Cole expected to advise the trustee for the debenture holders that afternoon that their security was in jeopardy and would be asked to postpone any action until after the company's proposed meeting with Dan Haughton which had been scheduled for February 2nd.

115. **On February 1st, 1971,** the main board meeting was attended by key HMG representatives and reported if there were no resolution of the situation via talks with Dan Haughton the next day, then HMG would be making a statement which would be preceded by a company statement and that would refer to a request to the Trustee for the debenture stockholders to appoint a receiver. Payments to creditors were now stopped. The board expressed surprise that the Trustee for the debenture holders had already been informed of the position Lord Poole said, by the Minister of Aviation and Supply. That afternoon Lord Cole, Ian Morrow and P. Benham went to the GREA offices with a view to handing in a copy of the letter drafted that morning. They were stopped at the door by a message from Prime Minister Edward Heath that he was still 'vitally interested and didn't want any irrevocable step to be taken'.

The Prime Minister had a telephone call with Dr H. Kissinger, Assistant to the President for National Security affairs, (the President being unavailable) informing him in confidence of the position Rolls-Royce found itself and the threat to the RB211 engine.

116. **On February 2nd, 1971.** A Lockheed team headed by Dan Haughton arrived in London for a joint programme review with Rolls-Royce management, Lord Cole, Ian Morrow and Hugh Conway, on the status of engines for the TriStar. Dan Haughton was reported as being 'buoyant' as to the progress with Lockheed's own financial problems. When informed over lunch that the British government had withdrawn financial backing and Rolls-Royce was now contemplating probable bankruptcy, Dan Haughton appeared dazed and unable to believe HMG would allow Rolls-Royce to become bankrupt. He was told that Lockheed would have to relinquish all claims under the existing contract and to accept a delay of 12 months on the delivery of production engines and an increase in engine price. A meeting was arranged later that day at 4:30pm with the Minister of Aviation and officials. Lockheed were informed that to save Rolls-Royce, there would need to be an increase in the RB211 engine price – including those already contracted for, of £100,000, plus an indemnity against any claims by Lockheed regarding

late deliveries. The serious cash requirements Rolls-Royce needed to continue involving between £100 million and £120 million were also outlined.

The price increase was not accepted at that stage, thought negotiable and negotiations on future engine prices would be possible; an indemnity against any claims by Lockheed in respect of late deliveries, deemed open to solution and a cash injection from HMG either supported or owed. The Ministry had earlier stated its position that it didn't believe the department could advance the sum of £10 million for work delivered and owed in the present circumstances described as a 'revolving door of credit' or make payments on account; Neither could it settle monies, in time relating to the Spey engine for the Phantom aircraft, as to be involved may make HMG liable to creditors and be in contravention of S.332 of the Companies Act 1948.

117. **On February 3rd, 1971,** representatives of the two companies met again. Lockheed said it needed more time to consider the matter before it could give any reassurances regarding increased prices or abandonment of claims. Any financial help would require an Act of Congress. The matter of claims from airlines remained a risk.

The Rolls-Royce Board met and was informed by Lord Cole, via Sir Henry Benson, that Mr. Packard, US Deputy Secretary of Defence had informed his President of the position. The President had requested more time so he could speak to the Prime Minister that afternoon. Dan Haughton was already at No 10 Downing Street, and he had indicated that he hadn't the authority or Lockheed the resources to make concessions to the contract terms, a delay in delivery or indemnify Rolls-Royce against any claims for delay or pay a price increase of £100,000 per engine.

Technically Rolls-Royce was by this time insolvent. None of the additional finance of £60 million negotiated in November 1970 was drawn on but even this amount would be insufficient to have any bearing at this point.

A statement to the Trustees was deferred. A full Cabinet meeting was held between 4pm-7pm. The Cabinet accepted that a receiver had to be appointed and hoped some arrangements could be made with the receiver for the RB211 to continue in some form and re-negotiate the contract with Lockheed.

A telephone discussion took place between the Prime Minister and the US President explaining both the Legal and HMG's position, that there was no alternative but receivership. The President said that Mr. Packard had kept him informed and that Mr. Packard was doing all he could to assist Lockheed. It was hoped the statement by HMG would indicate both Governments were working in close contact with one another.

118. **On February 4th, 1971**, Ernest Eltis sent a telex to the directors who were meeting at Board HQ in London. The telex sent confirmed that, following an overnight, full performance test run, engine 10011 of 'batch 3', after undergoing modifications and testing had achieved a thrust of 38,500 lb. at 1475°k on a standard day, equivalent to 34,500 lb. on an 84° day (even higher operating temperatures were also achieved during this run). The commitment to Lockheed was for a thrust of 38,000 lb. at 1475k on a standard day by the 12th of February. The contract provided for a thrust of 40,600 lb. on an 84° day. The performance was sufficient to have continued with the flight test programme. Whilst it would not have affected the decision of the board it would have a bearing on the continuation of the programme after receivership, indicating the engine could be a technical success.

119. **On February 4th, 1971**, at 08.45am, the Rolls-Royce Board met to make an announcement that, subject to talks with Lockheed and HMG, the RB211 programme 'would be discontinued, should be regarded as being suspended for the time being'. A letter to the trustee for debenture holders asking that a receiver be appointed for Rolls-Royce Limited was approved by the main board and it was also requested that dealing of the company's shares should be suspended and that a letter should be sent to the Chairman of the Stock Exchange forthwith.

The Company was effectively in receivership, and it would now be left to the newly appointed receiver (E. Rupert Nicholson) to take control of the company and secure its assets.

CHAPTER 7

FEBRUARY 4TH 1971, FALL OF A GIANT: ROLLS-ROYCE IS BANKRUPT!

Main Board approved statement on the Appointment of a Receiver published 4th February 1971.

"1. The Programme of the RB211 engine for the Lockheed Trijet aircraft, due in service in November this year has met with even greater problems than were first foreseen in November last.

2. The consequences can be summarised as follows: -

(i) While the board is confident that an engine fully capable of meeting the specification can be developed, the time required to achieve this will stretch beyond the time available under the contract with the Lockheed Corporation.

(ii) The costs of launching the engine will exceed by a wide margin the earlier estimates on which the company sought additional facilities of £60 million last November. None of the money has been paid; its availability from the Government and Banks was made conditional on a satisfactory report by independent accountants whose work has not yet been completed.

(iii) The earlier estimates of production losses are suspected to be substantially increased.

3. These factors give rise to additional costs and liabilities which are wholly beyond the financial resources available to Rolls-Royce. In these circumstances it is not feasible for them to proceed with the RB211 under the present contract.

4. The loss of resources already committed to the project combined with the losses which will arise on termination are on such a scale that they are likely to exceed the net tangible assets of the Company.

5. In the light of this situation the board have decided that they have no alternative but to ask the Trustees of the Debenture holders to appoint a Receiver and Manager.

6. The Government have been fully informed.

7. Rolls-Royce is engaged on important projects affecting the Government's defence programme; in addition, there is a need to keep in service Commonwealth and foreign air forces and a large number of civil and commercial operations; there are also important joint venture projects in progress with certain companies in America and certain companies in Europe. The board have had discussions with the Government as to the best possible means of protecting these interests and it is understood that the Government will be making an announcement later today.

8. Whatever arrangements will be made it is unfortunately clear that substantial redundancies are inevitable. The receiver will no doubt arrange early discussions with staff and unions. The board is distressed by the hardship that will be caused to many.

9. The board deeply regret the loss and embarrassment which will result from the failure of the company to meets its obligations to the shareholders, to the Lockheed Corporation and to banking institutions and trade suppliers.

10. The board consider that the shareholders should be given an opportunity of having an independent enquiry made into the facts of the situation which has arisen. It has therefore been resolved to convene an Extraordinary General meeting of the company to be held as soon as practicable at which the members will be asked to pass a Special Resolution in accordance with Section 165 (a) of the Companies Act 1948 requesting the Secretary of State for Trade and Industry to appoint one or more Inspectors to investigate the affairs of the Company and make a report thereon to the Secretary of State".

The company secretary had been asked to ensure that the Chairman of the Stock Exchange and the quotations department received a copy of the statement by 9:30 am. Board approval was obtained to ask the Chairman of the Stock

Exchange that dealings with the company shares should be suspended and that a letter to that effect should also be sent immediately. A letter was also sent to the Trustee for the debenture holders asking for a receiver to be appointed."

Later that day the Trustee appointed Mr. E. Rupert Nicholson as Receiver for Rolls-Royce Limited.

R. A MacCrindle, QC and P. Godfrey, FCA would later be appointed by the Secretary of State for Trade and Industry on April 23rd, 1971, to investigate the affairs of Rolls-Royce Limited.

THE APPOINTMENT AND WORK OF THE RECEIVER, E. RUPERT NICHOLSON

> *"You are not justice sitting on a monument with a set of scales; your only job is to get in the assets."*
>
> *Milner Holland, QC, Barrister*

From a conversation recalled in Rupert Nicholson's book, *Rolls-Royce Limited: Recollections of the Receiver and Manager, KPMG Peat Marwick, 1993 (pg. 28)*

At the age of 62, in early February 1971 and at a time of life when most people would be thinking about their forthcoming retirement, Rupert Nicholson was about to undertake probably his most challenging career assignment yet. A seasoned and highly experienced accountant, knowledgeable in all matters of insolvency and liquidation, Rupert had followed in his father's footsteps professionally since joining Peat, Marwick and Mitchell (PMM) as an insolvency practitioner. As a senior manager and later a partner in KPMG's (formerly PMM) insolvency team no one was more surprised than he was to receive a call from the Managing Director of Guardian Royal Exchange Assurance (GREA) regarding an urgent meeting he wanted him to attend about an impending receivership assignment.

Of course, that meeting held on February 1st, 1971, concerned none other than managing the receivership undertakings for Rolls-Royce Limited. It had been made clear from a previous meeting on the afternoon of the January 1st between Lord Cole and Ernest Bigland, Managing Director of GREA and trustee for the debenture holders, that their security was in jeopardy and, as would be the correct protocol, invited him to appoint a receiver. However, the Trustee was asked to postpone any action until after the Company's meeting with Dan Haughton on the 2nd of February. At that time GREA was the major debenture holder, with approximately £55 million committed to Rolls-Royce in the form of secured fixed interest loans, giving them first claim on the assets of Rolls-Royce Limited in the event of the company going into receivership.

Speculation had been mounting in the press for many weeks that a major company was in trouble and Rolls-Royce's fate finally was sealed by the appointment of the Receiver, to manage the receivership by GREA.

There was an air of unreality and disbelief permeating not only from the senior management and directors of Rolls-Royce but also from the citizens of the United Kingdom, many of whom read in astonishment that this could happen to one of their most distinguished and prestigious blue-chip companies whose name was always a byword for engineering excellence.

Events gathered at pace and Rupert Nicholson quickly had to form a receivership team to manage and run a multi-site, divisional operation. The team would eventually grow at its height to around 80 people. It comprised the brightest and most experienced staff and partners from within PMM.

Documented at the very beginning of Rupert Nicholson's recollections, that the press had been full of Rolls-Royce's troubles in January 1971and thought '*like many others that R-R had been unfairly the victim of changing political winds.*'

Having regard for the company as '*highly efficient technically*', he had however, an abiding memory of watching a television interview, given by Sir Denning Pearson, the Rolls-Royce Chairman, in which he reported Sir Denning as seeming to say, "*The duty of a company was to concentrate on technical excellence and leave profits to come, as they would, as a sure reward*". Ibid *(E.R Nicholson 1993)*

Rupert Nicholson wasn't particularly impressed with this statement which embraced the Rolls-Royce mantra and ethos of 'the pursuit of excellence', which was embedded in the company culture; indicating, if you produce an excellent product it will sell without the need to consider the cost involved. Although this would be challenged by the rapidly escalating development costs of the RB211, conversely, the Rolls-Royce Motor car division though equally of the same ethos, producing cars of the highest quality, could generate a profit, regardless of any increase in the product cost.

The fortunate factor in this crisis for the nation and the town, as it was in 1971, of Derby, was that Rolls-Royce was, even then, a diverse company, with a variety of engines within different market sector applications and it had both civil and military aero engine business contracts. The government therefore had to move very quickly to secure the continuity of supply of spare parts and engines of Rolls-Royce's military businesses that were critical to the defence of the United Kingdom. It would indicate its position to acquire assets of the aero-engine and industrial and marine turbine divisions as soon as the appointment of a receiver was taken. On the civil side there were already several mature engine programmes ongoing including the Spey, Tyne, Dart and Conway engines.

There was no doubt that quick decisions were required from the receiver if the RB211 project was to continue and survive, given that Rolls-Royce now found its very own future under threat and in jeopardy primarily due to the escalating costs of the RB211 project. Thankfully, the government gave the receiver authority to maintain the capability to produce the RB211 engine (research, development and testing work) until a time that the future of Rolls-Royce as a company could be decided. Then began the start of a constant round of shuttle negotiations between Rolls-Royce, Lockheed and the UK/US governments, involving the Lockheed Chairman, Dan Haughton, the receiver and HMG representatives, to secure the RB211 aero- engine contract and the futures of Rolls-Royce and Lockheed as companies.

The complexity of the lines of internal and external communication between the receiver and interested parties in the receivership as well as HMG departments is probably better shown by way of an Interface chart set out overleaf.

ROLLS-ROYCE MOTOR CAR DIVISION (MCD)

A brief overview

Rolls-Royce motor cars had always been considered the first and last word in car engineering excellence, producing instantly recognisable motor vehicles of quality and prestige since 1906. Based in Crewe, MCD was one of the first assets that the receiver looked at to sell following the bankruptcy. Interestingly at that point, MCD only accounted for over 5% but less than 10 % of Rolls-Royce's annual total business income. However, MCD was profitable and was one of the few, if not the only, independent luxury car maker, doing so. The company had a strong order book and held approximately £4 million in liquid assets.

The assets of the old company, Rolls-Royce Ltd., were being sold, but of course, these did not include the massive RB211 liabilities which the government had distanced itself from and which were left to be dealt with by the receiver. The separation of the Aero- Engine Division from MCD was believed to have been influenced by the government which was not interested in maintaining the car production business.

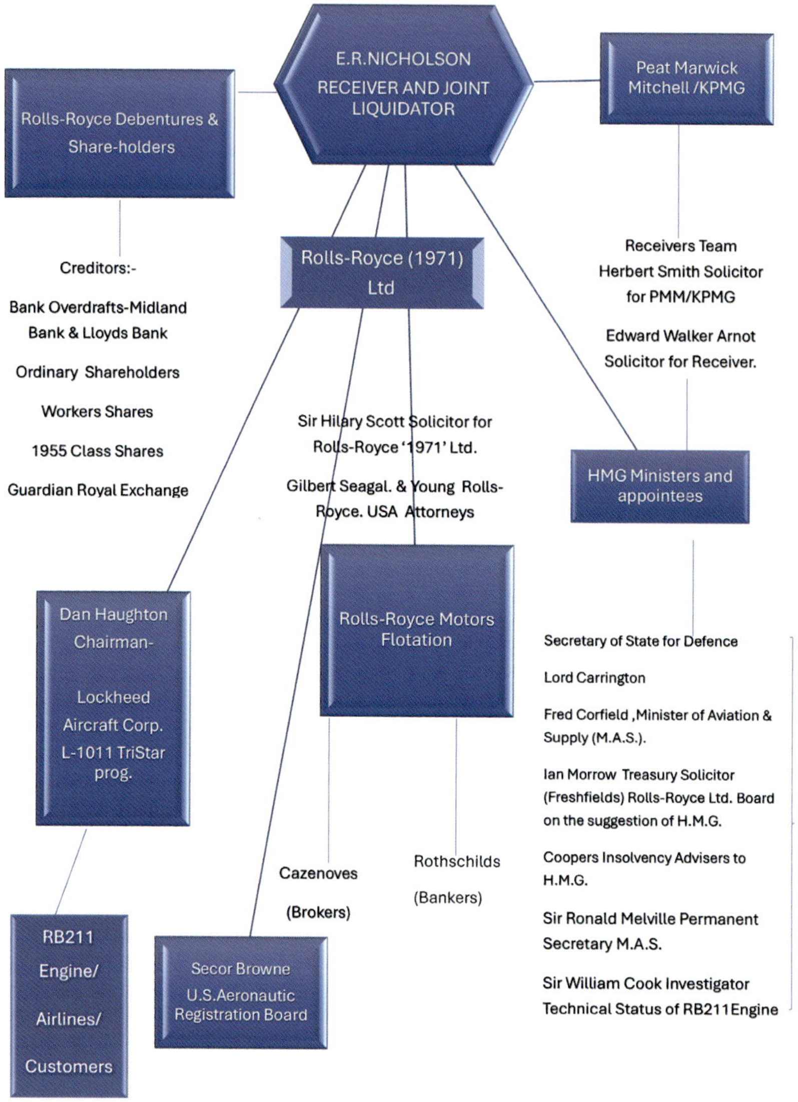

Matrix of receivership interfaces

Within three months of the bankruptcy, the receiver was instrumental in the formation of a new company, Rolls-Royce Motors Ltd., being incorporated on April 25th, 1971.

The new company had been formed from the Motor Car and the Diesel Divisions, previously the Oil Engine Division, as well as incorporating coachwork (made previously by Mulliner Park Ward) and the light aircraft engine interests of the old company.

The receiver had to assess whether to sell the company to a single party or to float the company on the stock exchange, which is what eventually occurred, all tender bids having been rejected, (the 'tender technique' having been used by the receiver before to optimise the proceeds of a share issue). Ibid (*Nicholson, E.R., 1993)*.

The receiver argued that Rolls-Royce motors be allowed to utilise the tax losses of the old parent company, which was rejected by the Inland Revenue and eventually on appeal, which delayed the flotation for about a year.

In the interim period, the government would use the liquid assets realised to continue the activities of the Aero-Engine, Industrial and Marine, and Small Engine Divisions that were so important to national defence and government contracts.

There was great confidence that the flotation of the company would show adequate, immediate and longer-term returns on investment. (*Financial Times 17.9.71)*

The turnover expected that year would be £34 million and employing 8,000 people at its Crewe car plant and in the Diesel Division, based at Shrewsbury, the Financial Times also ranked it in the UK's 300 largest companies with an asset value of £20 million and £34 million in sales. Guy Parsons had been asked to join the team by the Receiver along with another of his righthand men, Freddie McWhirter, to look after R-R Motors under the chairmanship of Doc Llewellyn Smith. As can be read in Guys letter to the author, (see images 58) following on from the enquiry by Rupert Nicholson as to whether he could look after the Motor Car Division for him, he was on the train to Crewe within the hour.

The public flotation was awaited eagerly and was expected to occur in 1972. But it would not be until May 1973, when the business climate was deemed right, and the appeal process exhausted regarding the tax issues, that Rolls-Royce Motors Ltd. was finally sold to Rolls-Royce Motors Holdings Ltd. in preparation for its flotation on the stock market, eventually realising £37.2 million for the liquidation account, almost twice what was expected and ultimately added to the payment to creditors and shareholders.

GUY PARSONS OBE – his personal history

Following Royal Navy service in World War II, Guy proceeded to take up a rewarding career with a London accountancy firm, Peat, Marwick and Mitchell (PMM), later KPMG, qualifying as an accountant in 1952. What followed was a career spanning over 42 years as a respected and astute insolvency partner and someone to whom floundering companies looked for sound advice.

Apart from Gloster Aircraft and Hawker Aircraft another one of his most memorable clients was the airline British Eagle, a small, short-haul airline that, after years of financial troubles, eventually went into bankruptcy and PMM were appointed as liquidators. Guy was the Manager of Operations during the process of administration. They had redundant BAC1-11 aircraft that Guy sold to a small group of British Eagle engineers and pilots which eventually became the company known as Monarch Airlines, which ceased operating in 2017.

As previously mentioned, Guy's reputation as an excellent negotiator was legendary and the British government (Ministry of Overseas Development) appointed him to negotiate and resolve sensitive international situations and his success rate was not to be underrated. After the Suez Canal war, the British government struck a deal with Egypt's President Nasser under which Britain would receive compensation for what had been lost by British companies operating in the Suez zone. Guy was asked to go over to Egypt several times to determine how much was owed and to see it settled.

In due course, he was appointed as a joint liquidator of the old Rolls-Royce Ltd company.[14]

It was acknowledged very early on that it was in the joint interests of the receiver and those involved with the motor company that it should remain independent. (Ibid *Financial Times 17.9.71).* Many other marques had been incorporated by other motor manufacturers such as Ferrari by Fiat. Rolls-Royce Motors was permitted use of the various Rolls-Royce trademarks and marque which was very precisely defined and controlled for future use and application. David Plastow, managing director, showed his gratitude and that of his colleagues in a letter (included on page 194) as testament to Guy's successful contribution to the company and overseeing the sale and flotation of Rolls-Royce Motors and maintaining its independence. (see also Image 78) His character and personality were credited for saving the company's morale.

14 A liquidator is appointed by the court to an insolvent company to collect the assets (after the receiver has completed his work on behalf of the secured creditors such as the debenture holders). A liquidator can review the conduct of the receiver to see if he should attack the receivership in any particular way.

Some of the models of Rolls-Royce cars that were being manufactured at that time included the Silver Shadow and the new Corniche. Worldwide demand remained strong for these products echoing the mission statement made by David Plastow, the managing director, "To make refined, elegant and long-lasting cars and not to be diverted into producing something smaller".

When the Motor Car Division was about to launch the 'Corniche'. Guy drove the new car home for a weekend trial, to the delight of his family and friends.

GILES CATTERMOLE (Former Crewe Graduate, Rolls-Royce Motor Car Division)

Recollections of working at Rolls-Royce Limited, Motor Car Division, Crewe

> *"I joined Rolls-Royce at Crewe (Motor Car Division) in September 1969, fresh out of a Mechanical Engineering degree. There were about a dozen of us graduate apprentices and all embarked on a two-year apprenticeship that would hopefully lead to a permanent job at Crewe.*
>
> *The structure was laid out for us on the first day:*
>
> - *Year 1: six, two-month attachments to as wide a variety of departments as possible, initially chosen for each of us by the Training Department, which was housed in temporary offices to the north of the Management Block.*
> - *Year 2: two, six-month attachments to those departments which, by mutual agreement, we were most likely to join.*
>
> *At the end of each attachment, our performance would be assessed and recorded, and we would be asked to assess the training we'd had.*
>
> *Morale was high; the parent company was doing well, but then rumours started to circulate about a potential sale of Rolls-Royce Motors, and we were told that there was to be no new design work on a replacement for the Silver Shadow model–already nearly five years old and starting to look somewhat outdated. Rolls-Royce Crewe was an old-fashioned company. When I was working in the Design Office (which was on the upper floor) there was a difficult class relationship between the designers there and the technicians downstairs in the Detailing Office. I recall one occasion when the Chief Designer decided he wanted to see the results of a design 'modification' he had introduced, and its effects on production operations. He went*

ROLLS-ROYCE MOTORS

The Managing Director

C. T. E. Parsons, Esq.,
Peat, Marwick, Mitchell & Co
11 Ironmonger Lane
London EC2P 2AR

17 May 1973

My dear Guy,

Thank you for letting me have a copy of your counter-inflation memorandum which I will mark, learn and inwardly digest.

In all of the frantic activity of the last few weeks I do not feel that I have properly expressed the enormous gratitude my colleagues and I feel towards you personally for all you did to produce the outcome of independence for our new Company.

There are many aspects of our relationship which I will remember for a very long time. Your sensitive handling of myself and my immediate colleagues at times of personal stress was quite masterly and always produced the results. Your grasp of the many factors in the situation with particular regard to taxation was brilliant, whilst your stage management of all the difficult personalities should be the basis of up-dating that standard work "How to win friends and influence people"!

You do indeed have many friends in the Company you helped to create and we look forward to being able to express the depth of our gratitude to you at a small dinner party we hope to arrange within the next few weeks.

Yours sincerely,

David

Letter of thanks to Guy Parson

downstairs to the front door of the design building and had to ask the way to the right part of the production line.

The first two, two-month assignments went well, with a well-structured programme at each department, good mentoring, and evaluation at the end. The first was a quality control job: looking at a track-rod end-cone fitting that was pulling through excessively on the nut being torqued up. The second was on a team trying to design a Silver Lady bonnet mascot that would meet US regulations for not damaging the head of any pedestrian the car ran into. Lots of fun, working with a macabre 'guillotine' that dropped a cradle of the weight of a human head onto a horizontal anvil holding the trial mascot.

We then heard about a BOAC VC10 aircraft, flying the London–Cape Town route as a flying test bed for the RB211 engine [with the RB211 replacing two Conway engines on the port side], which suffered a bird strike somewhere over Kenya. Neither the bird, nor the engine, nor Rolls-Royce, came out of it very well. The Hyfil carbon fibre fan blades started to delaminate in flight. It soon became clear, when this was repeated on a Hucknall test bed, that the fan blades would have to be replaced with titanium blades, at huge additional cost to Rolls-Royce. The parent company was in financial trouble and costs everywhere had to be cut to the bone and that included at Crewe.

When we went to the Training Department for our third assignments, there were just two secretaries frantically trying to match available assignments to apprentices. But it got sorted out in the end. I was given an assignment to design an automatic system to lock the doors once the car's speed had first risen above 10 mph. Many of the cars sold in the US had been bought by doctors. It was reported that drug addicts knew this, and would lurk at traffic lights and, once the Rolls-Royce had stopped, would open the driver's door and threaten the doctor, to get him to hand over his drugs bag.

At the end of the third assignment, we went to the Training Department to find the offices empty, the doors flapping open and the training files–including our prized official training logs for the I.Mech.E. certification–lying about on the vacant desks. We ended up roaming round the Crewe works site like feral dogs, looking for interesting departments and projects, and offering our services 'freelance', on the basis that Training was paying our salaries, so the department wouldn't have any on-costs.

I managed to get onto the diesel wankel project: to develop a very compact 900 brake horsepower (bhp) diesel engine for the UK's next-generation tank being developed by Vickers. This was fascinating cutting-edge stuff. Each 400x500x150 twin-rotor bank unit generated 160 bhp on the test bed on the first full-power run. So, a six-bank engine should have generated 1000 bhp, but–a bit too cutting-edge. We couldn't get the stator chrome plating round the spark plug to stand up for long enough, before starting to peel–with disastrous results. And the rotor tip seals didn't last long enough. I stayed with that project as long as I could, but by the time it was wound down, the situation at Crewe was deteriorating fast. It was clear that we were going to be sold. And without a new model on the drawing board, it would not be as a self-sustaining unit. Good people were looking for jobs elsewhere in other Rolls-Royce divisions. And the 'core' Rolls-Royce units were hovering like vultures looking for talent. I was offered a job at Ansty working with helicopter engines.

It was not just Rolls-Royce but all of England that was in a bad way in 1970 and I felt that designing systems for rich American doctors who were too lazy to flip the door lock switch was not a meaningful contribution towards a better world. So, I asked to be given two years leave of absence, to work as a U.N. volunteer in the developing world."

SUMMARY OF KEY EVENTS IN THE ROLLS-ROYCE RECEIVERSHIP PERIOD FROM FEBRUARY 1971 TO OCTOBER 1971

A wider perspective and understanding of the sequence of events from February 1971 that took place is given below by way of a condensed chronology. This includes some of the key events that occurred between the major interested parties: Rolls-Royce, Lockheed, the US government, the UK government, banks, airlines and the receiver. Information compiled in this chapter comes predominantly from E. Rupert Nicholson's book, Ibid *(1993)*. Additional material was also complied from the full 'Report to Government', Ibid MacCrindle R.A., QC and Godfrey P., FCA,*(1973)*; *Hansard:- HC Deb 4th February 1971, (Vol 810, c1922); HC Deb 8th February, 1971, (Vol 811 cc102-07); HC Deb 23rd February 1971, Vol 812 cc11-2W; Hayward, K. (1983); Pugh, P. (2001); Ruffles, P. (2014)*. Full acknowledgement is given to all sources.

1. **February 1st, 1971,** the Managing Director of GREA approached Rupert Nicholson in confidence, informing him that he needed to appoint a receiver of Rolls-Royce and wanted to know if he 'could and would' take on the role.

2. **February 4th, 1971,** Rolls-Royce filed for bankruptcy, citing losses on the RB211 engine development, and requested the appointment of a receiver. It announced that it was not feasible to proceed with its RB211 engine contract with Lockheed.

3. **February 4th, 1971,** following a telephone call, Rupert Nicholson and Graham Corbett attend the offices of GREA where they were told of a resolution passed by the directors requiring them to appoint a receiver as they were satisfied that their security was in jeopardy.

4. **February 4th, 1971,** Frederick Corfield outlined in Parliament the Governments proposal, to nationalise those parts of Rolls-Royce which would safeguard contracts important to national defence and collaborative projects with other countries. Portions of Rolls-Royce were subsequently to be reorganised as a government owned company Rolls-Royce (1971) Ltd., with Lord Cole as Chairman.

5. **February 5th, 1971,** HMG requested that the receiver maintain key supplies and spare parts to government departments and HMG military contracts. However, HMG did not accept any liability with respect to the RB211 engine contract.

6. **February 8th, 1971,** Frederick Corfield acknowledged in a Parliamentary debate that it was regrettable that Lockheed were not 'given more notice of this calamity'.

7. **February 9th, 1971,** a meeting took place with Lord Carrington, the Deputy Minister of Aircraft Supplies, the Rolls-Royce Chairman, Lord Cole, and the receiver.

   Lord Carrington (described as brusque by the receiver) urged him to repudiate the RB211 contract before he could discuss buying the business. (Within their 'special powers', a receiver can instigate repudiation of a contract to avoid claims from unsecured creditors in order to sell the assets free from these potential charges). Lord Cole made a last-ditch attempt to persuade HMG to buy the Motor Car

Division as well as the Oil Engine Division and Gas Turbine Divisions. HMG did not want to become involved in the motor car industry and if it did nationalise Rolls-Royce, it saw its priority as being only to nationalise the key aspects of the aero-engine business, where the government and national airlines were the main customers and because the R&D required government financial support due to the long-time lag before profitability. HMG also signalled it was willing to give the receiver an indemnity for all additional costs which he would incur in keeping the RB211 project capability alive.

8. **February 9th, 1971,** the receiver met with Rolls Royce's US attorney, Phil Gilbert (of Gilbert, Seegal and Young), who insisted that the RB211 contract was not onerous i.e. with heavy legal responsibilities and Lockheed and the American airlines would be helpful and reasonable but warned of a claim for heavy damages if it were repudiated.

9. **February 10th, 1971.** Telephone conversation between the receiver, Rupert Nicholson and Dan Haughton, Lockheed Chairman, confirmed that, with HMG authority, Rolls-Royce was 'maintaining capability to produce the RB211 engine' and was proceeding at full speed, in research, development and testing capability, at least until further decisions on the future of Rolls-Royce had been made. Dan Haughton confirmed he did not want the RB211 contract to be renounced. He stated he "wanted engines, not damages" and Lockheed would do all it could to help achieve this including supplying a large technical team if the contract was continued. The prospect of damages arising from Lockheed, or the airline companies had been a matter of concern.

10. **February 11th, 1971.** Good progress had been made on setting up the RB211 indemnity agreement, and supplies under government contracts which was renewed week by week. This only covered extra costs not overheads or depreciation. Work progressed on the RB211 as far as it could, considering all other factors.

11. **February 11th, 1971,** The Government announced interim financing to continue work on the RB211 engine pending the exploration of the possible continuation of the L-1011 programme.

12. **February 15th, 1971,** the receiver and Dan Haughton, visited the Coventry, Glasgow and Bristol sites meeting worker's representatives.

13. **February 15 & 16th, 1971,** the receiver met with Secor Browne, the chief of the American Aeronautics Registration Board sent over by US President Nixon on a fact-finding mission. He stated that, if necessary, they could adapt to the General Electric engine but would prefer the RB211 indicating even if the cost was higher than the contract price. His assessment of the RB211's high bypass ratio, three shaft design as well as fuel consumption and noise reduction were all favourable to the RB211 contrasted to the GE and P&W engines.
14. **February 16th, 1971,** there had been some resumption of share dealing in Rolls-Royce debentures and shares since February 8th and this had resulted in large numbers of buyers in America purchasing shares for only a few pence in value, whilst some shareholders in the UK had sold in order to establish a loss for capital gains purposes. The transfer of shares out of the country to the US was becoming untenable, jeopardising the position of Rolls-Royce as a British company. By the time the share register was closed, nearly 25 million shares had been purchased on behalf of buyers in the US by Guaranty Nominees Ltd. Efforts were underway to agree a facility to borrow £15 million of working capital from Lloyds Bank and Midland Bank, for Rolls-Royce to continue the RB211 business in the short term.
15. **February 18th, 1971,** William Cook was appointed by HMG to conduct a further independent technical assessment of the RB211 engine, current status and programme. He reported that the remaining RB211 technical issues were surmountable but, with a delivery date delay implication of six months. Dan Haughton was back in London again for initial conferences with Rolls-Royce and even though it had been renounced, made attempts to re-negotiate the contract, offering Lockheed's expertise and offering, according to the receiver in his recollections, Ibid., (*para.312*), to bail them out. He was adamant that he and his colleagues believed it preferable to get the RB211 right rather than switch to P&W or GE. He also had meetings with HMG officials regarding the terms for restoring the RB211 engine programme.
16. **February 20th, 1971,** discussions had commenced between the receiver, Lord Cole, Treasury solicitors and other interested parties about the valuation of the RB211 assets at a nominal value of £1.This, as also explained in the receiver's recollections, Ibid. (*para.506*), "*was to be assumed to make neither profit nor loss in calculating future profits for a going concern valuation of the business*". He went on to say this appeared "*naïve at first, but it was conceptual – the RB211 was a dead loss – so if Rolls-Royce (1971) Ltd. took it on it should not pay for it but neither should it deduct the losses it would certainly incur in the main*

contract, i.e. excluding extensions and spares". Following a meeting with the British Defence Secretary, Lord Carrington, and the Ministry of Aviation Supply, Dan Haughton returned to the US and announced he would resume talks with the British government representatives in early March.

17. **February 23rd, 1971**, Fred Corfield, Minister for Aviation, confirmed in the House of Commons that a new company, Rolls-Royce (1971) Limited, was to be formed to buy the aero-engine business assets of Rolls-Royce Ltd., from the receiver and would be registered that day.
18. **During February and March 1971,** Lockheed began exploring various means of continuing the TriStar L-1011 programme including investigating the availability of other US manufactured engines from GE and P&W. A re-run of the competition to supply the engine for the TriStar took place. Technical representatives from Lockheed's L-1011 customer airlines met in Burbank for several days and received detailed presentations from General Electric and Pratt & Whitney whilst a comparative analysis of the Rolls-Royce RB211 versus the Pratt & Whitney and General Electric engines was underway. Having compared these again, the Lockheed engineers found the RB211-22 superior and was selected. Furthermore, any change would be time-consuming, a possible delay of delivery to the airlines of 6 months and expensive, at a cost of $100 million according to Lockheed. As referred to in Peter Pugh's book, Ibid (2001, *pgs 230-231)*, apart from the thrust considerations, neither GE nor P&W would match the total British financial package. P&W accepted it had lost but GE came out fighting saying Government support for Lockheed should only be forthcoming if the company bought engines made in the US. However again, the US Government rejected this argument believing that Lockheed should buy the engines most suitable.
19. **Early March 1971,** a Heads of Agreement document was concluded with the Treasury, excluding some assets which Rolls-Royce (1971) Ltd. disposed of, including some subsidiaries and associate companies and some site workshop facilities.
20. **March 4th, 1971,** the DTI proposed a plan for a joint Lockheed–British company to carry out the engine programme (which didn't happen), agree an increased engine price for the RB211 along with mutual guarantees of programme security and a waiver of claims for penalties on the old contract . The British government would provide funding but required Lockheed to agree that the British government's investment would be repaid if Lockheed failed to carry through the L-1011

programme. Lockheed informed the government it could not accept elements of the British offer and indeed it could not raise the capital for a joint company. Dan Haughton would meet with L-1011 customers later in New York and present the results of Lockheed's evaluation of the availability of alternative General Electric and Pratt & Whitney Engines for the TriStar.

21. **March 13th, 1971,** Dan Haughton sent a message to Lord Carrington and reported that he had presented the British proposals to L-1011 customers, US government and banks. It was reported that the airlines also found the offer unacceptable, and they invited the British to present a 'best and final' offer for consideration by them and Lockheed.

    Pressure on Dan Haughton and the British government to conclude an early deal was assisted by Delta Airlines confirming its commitment to purchase 24 TriStars if the deal could reach a satisfactory conclusion.

22. **March 19th, 1971,** Talks with Lockheed and HMG had restarted in London with visits to the United States by a delegation led by Lord Carrington.

23. **On March 25th**, 1971, An agreement was reached whereby Rolls-Royce (1971) Ltd., would produce 555 engines with revised prices, delivery times, an amended price escalation formula for inflation and specification weight increase from 36,000lb to 38,500lb per ship set. (Ibid. Ruffles, P., 2014).

    The 555 engines would comprise 447 engines (149 shipsets i.e. 3 engines per aircraft) required by Lockheed and 108 engines required as spares by airlines. HMG would agree to finance Rolls-Royce with the development and production of the engine to enable it to maintain the necessary support facilities throughout the life of the engine as long as Lockheed would commit to continuing with the TriStar project and to produce the airframe. This required US Treasury and Administration support to arrange Government backing for an additional $250 million of bank loans.

    The receiver had taken part and agreed a tripartite deal allowing Rolls-Royce (1971) Ltd. to deduct $21 million from the price of the agreement in exchange for waivers for damages .

    The British and US Governments needed assurances that the airlines would still buy the TriStar, despite the delay and increase in price.

24. **April 8th and 14th, 1971,** Lockheed met with its lending banks (24 in total) to report on conditional agreements. Airline meetings also took place to discuss financing of the TriStar programme. Five days later, TriStar customers, banks and Lockheed conducted joint review meetings and reported that significant progress had been made.

25. **April 20th, 1971,** Prime Minister Heath said the British government would not proceed with the RB211 without guarantees as to the future of Lockheed. Six days later, the US banks provided additional loans to Lockheed, bringing the total bank financing to $400 million pledged, additional security for Lockheed.
26. **April 25th, 1971,** a separate company, Rolls-Royce Motors Ltd., was incorporated, formed from the Motor Car and the Diesel Divisions, – previously the Oil Engine Division.
27. **April 27th, 1971,** US Treasury Secretary, John Connally, told a Senate subcommittee that Lockheed would need a government loan guarantee.
28. **May 5th, 1971,** the American Administration announced its intention to seek congressional authorisation to guarantee loans to Lockheed.
29. **May 6th, 1971,** Secretary of the Treasury, Senator John Connally had a meeting with President Nixon at the White House and announced that the Administration would support a government loan guarantee and would send the legislation to Congress. Transworld Airlines announced one of the conditions that must be satisfied for it to continue with the L-1011, was the US government loan guarantee.
30. **May 10th, 1971,** the UK Parliament announced that the handover date of May 22nd, 1971, had been agreed between Rolls-Royce Ltd. and Rolls-Royce (1971) Ltd. and the re-negotiated agreement between Lockheed and Rolls-Royce (1971) Ltd. would be signed on that day but would still be subject to US Congress approval and ratification. Eastern Airlines signed a conditional agreement and reaffirmed its original purchase of 50 Lockheed L-1011s with Rolls-Royce RB211 engines.
31. **May 11th, 1971,** Dan Haughton, Ian Morrow and Lord Cole for Rolls-Royce presented a signed, conditional Lockheed/Rolls-Royce agreement to provide the RB211 engine for the TriStar at a new price and delivery times, subject to the fulfilment of conditions as to guarantees by the United States government of adequate finance and loan guarantees for Lockheed to ensure the L-1011 programme continued. Possible requirements for further development of the RB211 to power advanced longer-range versions of the TriStar were left for later discussions and decisions.
32. **May 13th, 1971,** the Nixon Administration sent a draft Emergency Loan Guarantee Act of 1971 to Congress for approval.
33. **May 22nd, 1971, the birth of Rolls-Royce (1971) Limited**

    The new company, Rolls-Royce (1971) Limited, was incorporated at midnight on May 22nd, 1971, and was formed to purchase substantially the whole of the undertakings and assets of the four original divisions of Rolls-Royce Limited previously connected with gas turbine engines.

34. **On June 10th, 1971**, the receiver announced that the debentures (£55 million) would be paid back in full, half being paid at the end of September 1971. Additionally, at least 50 pence in the pound would be paid to creditors following the announcement that the RB211 was proceeding.
35. **August 4th, 1971**, good news from the United States that the re-negotiated RB211 agreement had been agreed. The contracts were put in Escrow-left with a trusted third party, until they could be signed simultaneously.
36. **September 9th, 1971**, the contract came into force. A summit meeting was held at the Federal Reserve Bank in New York to finalise Lockheed's expanded bank credit agreement made possible by the Emergency Loan Guarantee Act. Attending were representatives of Lockheed's 24 lending banks, the US Treasury and Emergency Loan Guarantee Board, airline customers, Rolls-Royce representative Richard Turner who had been given authority to sign on behalf of Rolls-Royce, and the British government. Rupert Nicholson also took on the additional role of joint liquidator for Rolls-Royce Ltd.'s creditors.
37. **September 17th, 1971**, the voluntary liquidation meeting took place. In the background lots of meetings and discussions continued around the final valuation of the Rolls-Royce Limited assets. The work of the receivership team had been very challenging, and some very astute and shrewd management of the sale of assets had taken place over the course of the receivership. Over £21 million in interest was earned by these shrewd investments and was primarily attributable to the investment experience and skills of the appointed receivership banker, Gerry Tyrell.
38. **In mid-September 1971**, the receiver announced his intention to issue shares of Rolls-Royce Motors to the public, with the intention of having a full flotation in the summer of 1972 and accordingly Rothschild and Co. was appointed to handle this flotation. In March 1973, a prospectus for a sale tender bid was issued which eventually realised £37.2 million for the liquidation account.
39. **October 4th, 1971**, the incorporation that had occurred in May, enabled the old company, Rolls-Royce Limited, to be placed into liquidation on this date.

ROLLS-ROYCE MAIN BOARD MEMBERS AS AT FEBRUARY 4TH 1971

Showing dates of appointment and previous appointments where known

LORD COLE -CHAIRMAN appointed November 1st 1970.

SIR DENNING PEARSON -DEPUTY CHAIRMAN appointed November 1970. Previously Chairman and CEO 1969-1970.

IAN MORROW-DEPUTY CHAIRMAN & CHAIRMAN EXECUTIVE COMMITTEE appointed July 1970-1971.

WHITNEY STRAIGHT CBE, MC, DFC -DIRECTOR 1955-1971; EXECUTIVE VICE CHAIRMAN,1956-1957; EXECUTIVE DEPUTY CHAIRMAN 1957-1960; NON-EXECUTIVE DEPUTY CHAIRMAN 1960-1971.

D. PEPPER-DIRECTOR OF PERSONNEL

F. (Doc) LLEWELYN SMITH -GROUP MANAGER AUTOMOTIVE & SUBSIDIARIES. Previously Chairman of MDD & OED 1967.

H. G. CONWAY-GROUP MANAGING DIRECTOR GAS TURBINES appointment announced in the September 17th 1970-Brochure to shareholders. Previously Managing Director BED 1967.

GEOFF FAWN-MANAGING DIRECTOR DED. Previously Managing Director MDD, January1967.

R. NICHOLSON- RB211 PROGRAMME DIRECTOR. Previously Production Director January 1967.

H.E. TREVAN-HAWKE- FINANCIAL DIRECTOR. Previously Financial Controller, May 1967.

D. FLEMMING-DIRECTOR OF CENTRAL PLANNING GAS TURBINES, previously company adviser management systems 1968-1969; General manager DED,1969-1970, Director 1970-1971; Financial Director DED,1970.

<u>NON-EXECUTIVE DIRECTORS</u>

LORD BEECHING -appointed August 1970-1971

J. L. E. SMITH -appointed January 1967

SIR MAURICE LAING 1968-1971

F. T. HINKLEY, CBE, 1970-1971, previously Director DED 1958-1968; Commercial Director 1960-1967 & Acting Managing Director DED in 1968 & 1971.

54. Rolls-Royce Main Board of Directors as at February 4th, 1971.

55. Rolls-Royce receiver, Rupert Nicholson's portrait and treasured mementos from Rolls-Royce employees.

56. Plinth supporting the RB211 engine model as seen in the previous photograph. This was Rolls-Royce employees' gift to receiver Rupert Nicholson, for saving the company.

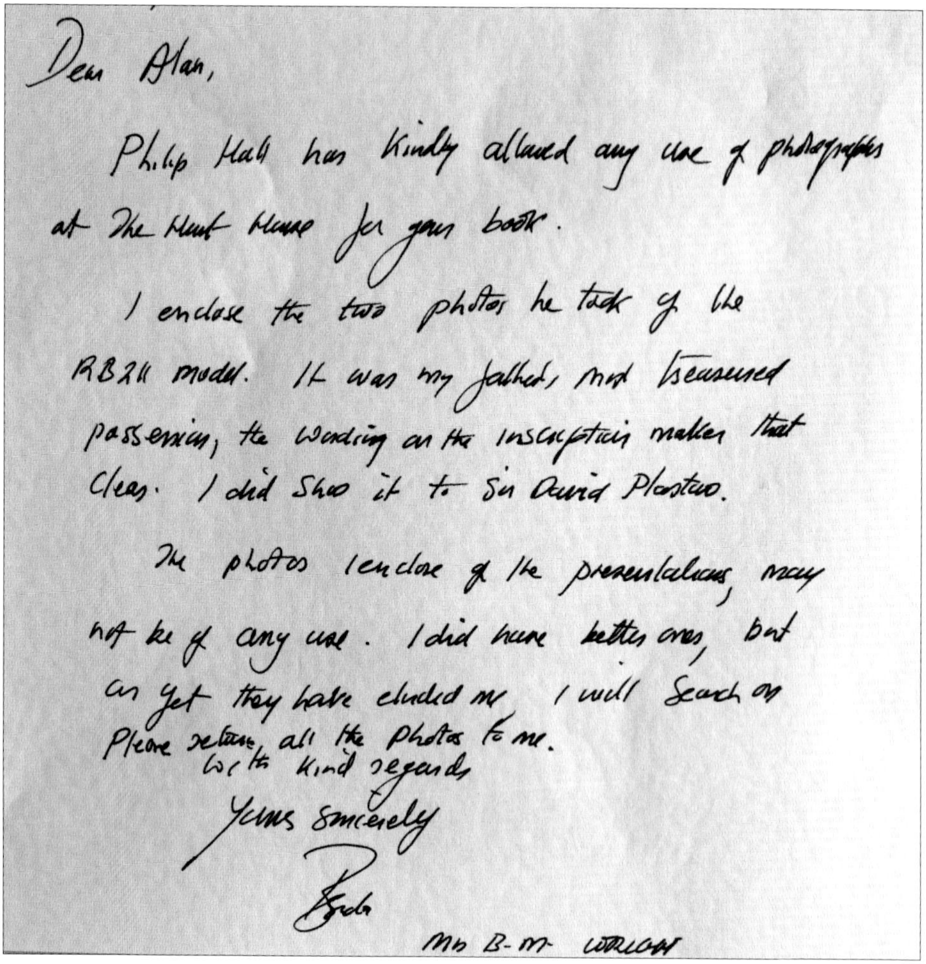

Dear Alan,

Philip Hall has kindly allowed any use of photographs at The Hunt House for your book.

I enclose the two photos he took of the RB211 model. It was my father's most treasured possession, the wording on the inscription makes that clear. I did show it to Sir David Plastow.

The photos I enclose of the presentations, may not be of any use. I did have better ones, but as yet they have eluded me I will search on.
Please return all the photos to me.

With kind regards

Yours sincerely

Brenda

Mrs B.M. Wright

57. *Letter to the author from Mrs. Brenda Wright, the receiver's daughter.*

Alan Edge

Guy Parsons O.B.E.

(1)

16/7/18.

Rolls. Royce

Dear Alan.

Following our talk today I am setting down some memories of the crash in February 1971.

At that time I was an Insolvency Partner at Peat Marwick and was part of Rupert Nicholson's team. He was a brilliant Insolvency Expert and he should have been Knighted. I believe he might have upset someone in high places.

I recall that one morning in the office I went into the Partner's toilet and Nick (as he was known) was also there. He told me he had just been appointed Receiver of RR and was I busy. If not could I look after the Motor Car Division for him. Within an hour or so I was on the train to Crewe

58. Guy Parsons, part of the receivership team and PMM liquidator–letter to the author of his receivership recollections.

(2)

Guy Parsons O.B.E.

I found the Motor Car division in quite good shape and I did the following

1. Appointed David Plastow as M.D.
2. Increased the no. of cars produced
3. Increased the Price of Cars.

At that time, they were about to launch a new open car – the Corniche and I had to approve this. However, I decided to take the car home for the week end and I was the envy of my friends – Obviously I approved it

Eventually we transferred the Motor Car Division into Rolls Royce Motors Ltd and we floated this company on the Stock Exchange. I believe that this was the first time a Receiver had ever done such a thing

58. page 2, continued.

(3)

Guy Parsons O.B.E.

After that the Company was acquired by Vickers and Sir David Plastow became Chairman.

The company changed hands a bit after that and is now owned by BMW. The cars are made at Goodwood and when I went to the Palace for my OBE they lent me a top-range car.

I attach a copy of my Biography and you will see references to RR on pages 10/11 and 12.

Please return this G

As you know I had nothing to do with the Aero Engine division which Nick sold to the Government for £1 and a deal was done with Lockheed over the RB 211 Engines.

58. page 3, continued.

(4)

Guy Parsons O.B.E.

When Nick retired I replaced him as Liquidator of RR but there was little to do after that

As far as I know, the liquidation has been completed but I am not too sure about that. Roger Oldfield would know

I hope this helps

Regards

Guy Parsons

58. page 4 continued.

59. View across Derby marketplace February 1971 – investors queuing to withdraw investments from Derbyshire Building Society, Iron Gate, Derby.

60. Rolls-Royce workers in London, February 1971.

61. Rolls-Royce workers– march to save jobs, Osmaston Road, Derby, February 1971.

62. Rolls-Royce workers– protest march to save the company, London, February 1971.

63. Rolls-Royce workers– protest march to save jobs, Full Street, Derby, February 1971.

64. Rolls-Royce employee at St. Pancras station – march to save jobs, London, February 1971.

65. Derby clergyman, the Rt. Rev. Cyril Bowles, and Derby Mayor, Miss E. Grimwood-Taylor sign the Rolls-Royce petition on saving jobs, with Charlie Hunt, Rolls-Royce Convenor seen left.

66. Derbyshire MPs Johnson, Rost and Whitehead – petition being handed in to No. 10, Downing Street, to save Rolls-Royce, February 1971.

67. Petition to save Rolls-Royce, Peter Rost, MP, third from the left, Westminster, February 1971.

ROLLS-ROYCE LIMITED

E R NICHOLSON - RECEIVER & MANAGER
APPOINTED 4TH FEBRUARY 1971

17th March 1971

K.P. Miller Esq.,
(Ch. No. 59706, Dept. 7380)

Dear Sir or Madam

You will be aware that due to the Company's financial position I was appointed on 4th February as Receiver and Manager. Since my appointment it has again been confirmed that the Derby Engine Division have passed the peak of effort required by their programmes in terms of staff employees and indirect hourly paid employees. The RB 211 project is proceeding for the time being with support from the Government. If the RB 211 project continues it will be on a new delayed programme which will not absorb the present substantial surpluses in staff and indirect manpower.

As I previously indicated in my notice of 5th February, it is therefore necessary for me to initiate a significant reduction in the number of employees. Consultations are taking place with the Trade Unions concerned, and in line with the procedures put forward, those to be declared redundant are being identified.

I am very sorry to have to tell you that you are one of those to be declared redundant. As Receiver and Manager, I am not bound by your contract of employment with the Company nor by any agreements between the Company and the Trade Unions. You will, however, be given 8 weeks notice from 19th March 1971. You will leave the Company's service on 26th March 1971. I have arranged that on that date you will be paid a sum equal to 7 weeks pay in lieu of the remainder of your notice and no tax will be deducted from this payment. You will also receive any payment in lieu of holidays to which you may be entitled, but this will be subject to income tax.

...2

P.O. Box 31, DERBY, DE2 8BJ
Telegrams: 'Roycar, Derby' Telex: 37645 Telephone: Derby 42424 Ext.

68. Rolls-Royce employee redundancy letter, March 17th, 1971.

17th March 1971

- 2 -

You are also entitled to receive redundancy payment in accordance with the Redundancy Payments Act 1965 if you are eligible for such payment under the terms of the Act. The Company is not making this payment, and therefore in accordance with Section 32 of the Act you should apply to the Department of Employment for payment of the amount due to you. Further information on the amount of and eligibility for this payment and the method and timing of your application to the Department of Employment are shown on the attachment to this letter.

To the extent that the arrangements set out above provide you with a smaller sum than you would have been entitled to under your terms of employment with the Company, you will be entitled to claim the difference in the same way as other unsecured creditors of the Company.

Your claim on the Company can include payments to which you are entitled in accordance with the Works or Staff Handbook, or in accordance with any collective agreement between the Company and a trade union of which you may be a member, or in accordance with your individual service contract if you hold one. How much of your claim would be met eventually will depend on the amount available to the unsecured creditors.

Further information on specific points is given in the attachment to this letter.

Yours faithfully

E R NICHOLSON
Receiver and Manager

68. page 2 continued.

GENERAL INFORMATION TO REDUNDANT EMPLOYEES

1) Redundancy Payments Act

a) A summary of the main conditions covering payments under the Redundancy Payments Act is enclosed for guidance in arriving at the payment to which you are entitled under the Act.

b) Arrangements will be made for you to claim the amount due to you and you will be advised by your local P & A officer of the time and place for signing the necessary form.

2) Company Pension Position

a) Under age 55 men (50 women)

The following options will be available to you if you are a member of the Company Pension Scheme and under age 55 (50 women) on 26th March 1971:

(i) You may take a return of your own contributions to the scheme (together with 3% compound interest) less deductions for tax and the State Graduated Pension Scheme, or

(ii) In certain circumstances you may leave your contributions in the Fund and receive a paid up pension at age 65 (60 for women), or

(iii) You may transfer the cash equivalent of your paid up pension to another Company's pension fund if that fund has the necessary transfer rules.

It is hoped to advise you of the net amount of refund of your Pension contributions and of the amount of your paid up Pension before you leave. If you have not been advised before you leave you will be contacted shortly after you leave by the Pensions Office.

b) Age 55 and over men (50 and over women)

The following options will be available to you if you are a member of the Company Pension Scheme and age 55 or over men (50 and over women) on 26th March 197

(i) (ii) and (iii) as in a) above, or

(iv) An immediate pension.

...2

69. Rolls-Royce redundancy letter pay terms, 1971.

- 2 -

It is hoped to advise you of the net amount of refund of your Pension contributions and of the amount of your immediate and paid up Pension before you leave. If you have not been advised before you leave you will be contacted shortly after you leave by the Pensions Office.

3) <u>National Insurance and Income Tax</u>

You will be given, on leaving, your National Insurance Card stamped up to the date of leaving together with your Income Tax Form P45. You will also be supplied with a copy of your Form P60 which is necessary for you if you later claim the Earnings Related Supplement from the Department of Employment (see 4 below).

4) <u>Unemployment Benefit</u>

Basic information on the unemployment benefit available (including Earnings Related Supplement) from the State is available from your local P & A officer. Enclosed is a letter of introduction to the Department of Employment for the purposes of registration for employment and benefit. The registration form is also enclosed and ~~this should be completed by you and returned before you~~ leave to the Department of Employment who have temporary offices at the main Company sites.

5) If you require any further information, would you please contact your P & A officer.

69. page 2, continued.

DEPARTMENT OF EMPLOYMENT

Employment Exchange
10 Normanton Road
DERBY

Telephone: Derby 46461

March 1971

I understand you will be leaving Rolls Royce and I am writing to let you know that my department has made special arrangements to help you find another job.

If you want our assistance will you please fill in the attached form ED 954 as fully as you can and take it at a time and to a point on your site which will be given to you either by means of notices or by your supervisor. There you will meet one of my officers who will explain what we can do to help you and answer any queries you may have. We are in touch with all employers in the area and most of those outside it and I hope that amongst the jobs we shall hear about will be one which will appeal to you.

J M SUTTON
Manager

70. Department of Employment letter sent to Rolls-Royce workers selected for redundancy, March 1971.

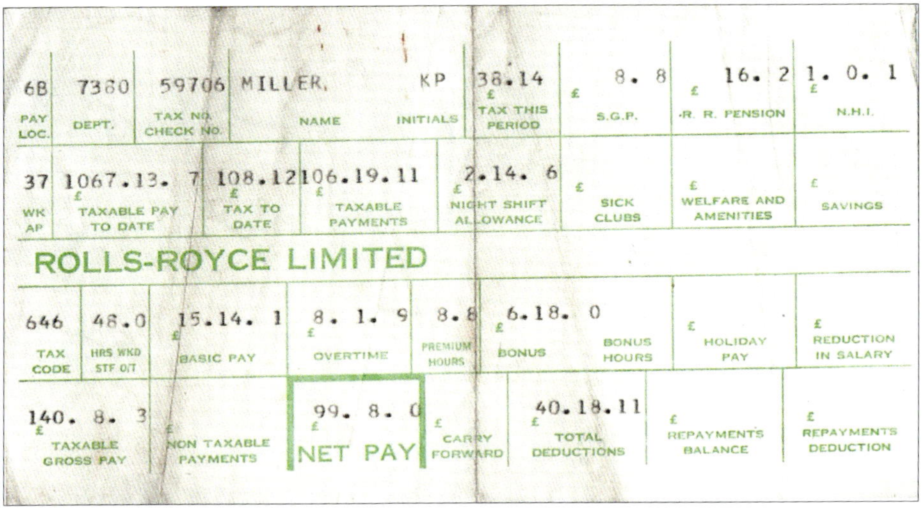

PAY LOC.	DEPT.	TAX NO. CHECK NO.	NAME	INITIALS	TAX THIS PERIOD	S.G.P.	R. R. PENSION	N.H.I.
6B	7360	59706	MILLER,	KP	£ 38.14	£ 8. 8	£ 16. 2	£ 1. 0. 1

WK AP	TAXABLE PAY TO DATE	TAX TO DATE	TAXABLE PAYMENTS	NIGHT SHIFT ALLOWANCE	SICK CLUBS	WELFARE AND AMENITIES	SAVINGS
37	£ 1067.13. 7	£ 108.12	£ 106.19.11	£ 2.14. 6	£	£	£

ROLLS-ROYCE LIMITED

TAX CODE	HRS WKD STF O/T	BASIC PAY	OVERTIME	PREMIUM HOURS	BONUS	BONUS HOURS	HOLIDAY PAY	REDUCTION IN SALARY
646	48.0	£ 15.14. 1	£ 8. 1. 9	8.8	£ 6.18. 0		£	£

TAXABLE GROSS PAY	NON TAXABLE PAYMENTS	NET PAY	CARRY FORWARD	TOTAL DEDUCTIONS	REPAYMENTS BALANCE	REPAYMENTS DEDUCTION
£ 140. 8. 3	£	£ 99. 8. 0	£	£ 40.18.11	£	£

71. Rolls-Royce employee's pay slip, 1971.

ROLLS-ROYCE LIMITED

(IN VOLUNTARY LIQUIDATION)

Receiver: E. R. NICHOLSON, F.C.A.

Estates House,
66 Gresham Street,
London, E.C.2.

Telephone: 01-606 8888.

18th November, 1971.

To: Stockholders and Creditors.

Dear Sir(s) or Madam,

At the Extraordinary General Meeting of the Company held on 4th October, 1971, the Resolution to place the Company in Creditors' Voluntary Liquidation was passed on a Poll by a substantial majority. The Meeting of Creditors held the same day approved the appointment of the following as joint liquidators of the Company:—

E. R. Nicholson, F.C.A. of Peat, Marwick, Mitchell & Co. (The Receiver of the Company)
Sir William Slimmings, C.A. of Thomson McLintock & Co.
K. D. Wickenden, F.C.A. of Thornton Baker & Co.

At the meetings the following were elected to serve on a Committee of Inspection to advise and assist the Liquidators:—

Creditors' Representatives

R. A. Clark, Vice-Chairman of Hill Samuel Group Limited.
A. J. Harrison, Executive Officer of Derby Chamber of Commerce.
R. M. Meatyard, F.C.A. Company Secretary of Associated Engineering Limited.
J. N. Prentice, F.C.A. Partner in Deloitte & Co.
J. W. Shield, C.A. Group Treasurer of Lucas Industries Limited.

Stockholders' Representatives

V. H. F. Hopkins, C.Eng., F.I.Mech.E., F.I.Ag.E., M.S.A.E. Consulting Engineer.
A. M. Mackintosh, C.A.
R. R. Masters, Assistant Investment Manager of the Prudential Assurance Company Limited.
D. R. McDonald, C.A. Late Financial Director and Secretary of Rolls-Royce Limited.
J. P. Waldron, Solicitor of Derby.

In the course of the Meetings, the following points of interest to both Stockholders and Creditors were announced:—

(1) **The Department of Trade and Industry Inspectors**
The Chairman stated that the Board of Directors were reinforced in their view that the appointment of the Inspectors was the only possible course under the circumstances, and that they were hard at work on their enquiries. They were making good progress, but it was impossible to estimate when their report would be completed.

(2) **Estimated Statement of Affairs** (copy enclosed)
The Receiver explained the Statement and made the following points:—

(*a*) The Statement is made up to 14th August, 1971, which was the latest practicable date, and therefore covered the period of the Receiver's trading. The Receiver had realised in the early days that the Debentures would be repaid in full and therefore his success or failure affected both the unsecured creditors and the members. In the event, the Statement shows an improvement in the overall position compared to the draft statement at 4th February, the date of the Receiver's appointment, which was published in June.

(*b*) Regarding the price to be paid for the assets transferred to Rolls-Royce (1971) Limited, the Receiver explained that it was not possible to estimate the amount as there was a danger that this would prejudice the negotiations. He emphasised that it was fundamental to a realisation of the true position that the negotiators on behalf of Rolls-Royce (1971) Limited maintained that Rolls-Royce Limited

72. Rolls-Royce Ltd. in voluntary liquidation. Notice to stockholders and creditors from the receiver dated November 18th, 1971.

ROLLS-ROYCE LIMITED

ESTIMATED STATEMENT OF AFFAIRS AT 14th AUGUST, 1971

	Book Value £ million	Book Value £ million	Estimated Realisable Value £ million	Estimated Realisable Value £ million
ASSETS				
Sold to Rolls-Royce (1971) Limited (Note 2 (*a*))				
Stock and Work in progress	92.0			
Property and Plant	76.4		Cannot yet be ascertained	
Subsidiary and Associated Companies	6.5			
		174.9		
Sold to Rolls-Royce Motors Limited (Note 2 (*b*))				
Stock and Work in progress	14.0			
Property and Plant	4.0		20.0	
Subsidiary and Associated Companies	1.2			
		19.2		20.0
Other Assets				
Cash (Note 2 (*c*) (i))	56.6			
Deduct: Payments to Debenture Holders	29.9			
	26.7		26.7	
Debtors (Note 2 (*c*) (ii))	51.1		51.1	
Property	0.7		2.0	
Subsidiary and Associated Companies	9.5		7.5	
		88.0		87.3
		282.1		
Deduct: Payment on Account of the Proceeds of the Sale to Rolls-Royce (1971) Limited		30.0		
Total Book Value (Note 3)		252.1		
Total estimated realisable value of assets, excluding the balance of the proceeds from the assets sold to Rolls-Royce (1971) Limited Carried forward				107.3

73. Rolls-Royce receiver's estimated statement of affairs at August 14th, 1971.

	£ million	£ million	£ million
Total estimated realisable value of assets, excluding the balance of the proceeds from the assets sold to Rolls-Royce (1971) Limited brought forward			107.3
Deduct (Note 1)			
Receiver's Creditors (Note 1)		9.8	
Preferential Creditors (Note 4)		10.9	
Debentures including accrued interest	56.6		
Less: Repaid on 15th September, 1971 (Note 5)	29.9		
		26.7	47.4
Estimated surplus available for Unsecured Creditors, subject to costs of realisation and excluding the balance of the proceeds from the assets sold to Rolls-Royce (1971) Limited			59.9
Deduct: Unsecured Creditors (Note 6)			
Trade		56.9	
Banks and Acceptance Houses		38.6	
Amounts due to Subsidiaries and Associates		.3	
Industrial Reorganisation Corporation		10.4	
Unsecured Loan Stock		1.8	
Other		7.8	
		115.8	
Add: Contingent creditors (say)		5.0	
			120.8
			(60.9)
Deduct: Share Capital (Note 7)			
Ordinary Stock		63.9	
Workers (1955) Stock		2.4	
			66.3
Estimated deficiency for stockholders, subject to costs of realisation and excluding the balance of the proceeds from the assets sold to Rolls-Royce (1971) Limited			(127.2)

THE ATTACHED NOTES FORM PART OF THIS STATEMENT

73. page 2, continued.

Notes to The Estimated Statement of Affairs

1. The Receivership

The Receiver and Manager, who was appointed on 4th February by the Trustees of the Debenture Holders, has dealt with the assets as shown in Note 2 below. The proceeds of the assets realised by him must be used, in order of priority, to pay:

(*a*) The Receiver's costs and trading liabilities

(*b*) The Preferential Creditors, and

(*c*) The Debentures, plus outstanding interest.

He was responsible for the management of the whole of the Company's undertaking until 22nd May when the four divisions comprising the Gas Turbine business were handed over to Rolls-Royce (1971) Limited; the rest of the assets are still either directly or indirectly under his control. During this period the Receiver and Manager incurred considerable liabilities (both contingent and otherwise) of which £9.8 million have been included in this Statement. It must be recognised that it is impracticable for the Receiver to hand over all the assets under his control while the contingent claims are outstanding.

2. Assets

(*a*) Rolls-Royce (1971) Limited has acquired the Gas Turbine business (excluding the debtors) on the basis of its value on a sale by a Receiver to a willing buyer. The negotiations to settle the price are about to begin, but in default of agreement between the parties the Sale Agreement provides for the value to be fixed by an independent assessor. Any estimate which might be made of the outcome of these negotiations, or of the reference to the assessor, would be liable to prejudice their outcome and cannot be given at this stage. Under the Sale Agreement, it was agreed that certain features of the transfer should be settled at a future date; these are the subject of continuing negotiations and it is not possible to estimate their impact upon the final price.

In the sale agreement, it is provided that the RB 211 project should be deemed, for valuation purposes, to make neither profit nor loss and the consideration for the assets relating exclusively to the RB 211 (estimated to amount to some £25M) is £1; however as a term of the settlement of these matters, this Company is relieved of most substantial claims which would otherwise have been aggregated with the present unsecured creditors.

The Receiver has already received £30 million on account of the sale price. Interest will be received on the balance of the proceeds.

(*b*) The Motor Car and Diesel Engine business has been sold to a new subsidiary, Rolls-Royce Motors Limited. The Receiver is planning to make a public issue (probably in early 1972). Again the giving of a realistic estimate of the proceeds might prejudice a satisfactory realisation, and the "realisable value" of £20 million is given as an arbitrary indication.

(*c*) *Other Assets*

(i) *Cash* This includes the £30 million received on account of the sale price for the assets sold to Rolls-Royce (1971) Limited, the proceeds of assets sold by the Receiver, and the return from the Receiver's trading.

(ii) *Debtors* A significant proportion of the outstanding debts are due under long term Government contracts. Provision would ordinarily have been made in the Company's accounts for claims and other adjustments but in reviewing these debts allowance has also to be made for offsets and attachments. It is not possible to separate these different matters and the Book Value of the debtors has been reduced as though all these provisions had been written into the books.

73. page 3, continued.

3. **Book Values**

(*a*) The Book Values shown are those at the following dates;

(i) Assets sold to Rolls-Royce (1971) Limited—31st March, 1971 (The date of the valuation for the purpose of computing the sale price).

(ii) Assets sold to Rolls-Royce Motors Limited—19th June, 1971 (The date of transfer).

(iii) Other Assets—14th August, 1971.

(*b*) Debtors—see note 2 (*c*) (ii) above.

4. **Preferential Creditors**

(*a*) The Receiver had paid Preferential Creditors amounting to £4.2 million by 14th August, 1971.

(*b*) The Banks have made claims in respect of advances for the payment of wages and salaries prior to the appointment of the Receiver. These claims are not expected to exceed £9.9 million and this sum has been included in the Statement.

5. **Debentures**

After the date at which this Statement is drawn up, the Trustees to the Debenture Holders announced the payment of all accrued interest and 50 per cent. of the capital value of the Debentures, and £29.9 million was duly paid.

6. **Unsecured Creditors**

There are sums due to and from subsidiaries of this company which are either included in the prices to be obtained for the assets sold to Rolls-Royce (1971) Limited or which, due to offsets, have no effect upon the estimated surplus available to the unsecured creditors and they have accordingly been omitted from this Statement.

In addition to the claims shown, there are various contingent and unquantified claims. These include damages for breach of contract and potential warranty claims. All claims so far known are being resisted; an arbitrary provision of £5 million has been included in the Statement, as a number are of material amount and some provision is essential.

7. **Share Capital**

The Workers (1955) Stock rank equally with the Ordinary Stock in winding-up.

0713 27187 B., M. & Co., Ltd. S5614.

73. page 4. continued.

In the Matter of ROLLS-ROYCE LIMITED

and

In the Matter of THE COMPANIES ACT 1948

NOTICE IS HEREBY GIVEN that the Creditors of the above named Company, which is being voluntarily wound up, are required on or before 31st December 1971 to send their names and addresses together with particulars of their debts and claims to the Joint Liquidators of the Company by addressing the same to the Joint Liquidators of Rolls-Royce Limited, P.O. Box 432, Estates House, 66 Gresham Street, London, E.C.2 or in default thereof they will be excluded from the benefit of any distribution made before such debts are proved.

Dated this 18th day of November, 1971.

Signed E. R. Nicholson
Sir William Slimmings
K. D. Wickenden
Joint Liquidators

Notes:—

1. Former employees of the Company are required to quantify the amount of their claims against the Company and are requested to state the division of the Company in which they were employed together with their works number (where appropriate).

2. The nature of each claim should be stated and, in particular, whether it is founded upon:—

(*a*) a breach of the Redundancy Agreement dated the 4th February 1964 made between the Company and the following Staff Unions:—
D.A.T.A.; A. Sc. W; C.A.W.U.; and A.S.S.E.T.

(*b*) a breach of the Redundancy Agreement between the Company and the Manual Workers' Unions

(*c*) a breach of some other Contract with the Company, in which case brief particulars of the nature of such Contract and the date thereof should be stated

(*d*) some other, and, if so what, obligation of the Company, whether statutory or otherwise, stating concisely the nature of such obligation.

3. Particulars should be given of the period in respect of which each claim relates together with details of any emoluments or payments received which would mitigate the claim against the Company and which must, therefore, be deducted in arriving at the total of such claim.

4. Any entitlement to priority under Section 319 of the Companies Act 1948 must be established when submitting claims and particulars of the nature of such title together with the amount to which it extends should be stated.

5. All claims must be submitted to the Joint Liquidators at P.O. Box 432, Estates House, 66 Gresham Street, London, E.C.2, but may subsequently be dealt with from the divisional locations.

6. Acknowledgments of claims will be sent on receipt, but creditors must expect that, because of the size and complexity of the liquidation, it may be some considerable time before they receive a further communication.

0111 29308 B., M. & Co., Ltd. S8386P

74. Rolls-Royce Ltd. notice to creditors under the Companies Act, dated November 18th, 1971.

75. First instalment of £5,000,000.00 to the receiver for the sale of assets of Rolls-Royce Motors, Diesel and Mulliner Park Ward Divisions.

76. Rolls-Royce management and unions visit Lockheed for productivity discussions.

77. Guy Parsons receiving his OBE from Prince Charles in recognition of his career and charitable work.

Guy Parsons: 'Rolls-Royce was rather special'

Roller relief

The Silver Lady, proud emblem on the world's most expensive motor car, has been the envy of millions for years. But as a symbolic representation of the spirit of ecstasy it looked decidedly dodgy in the early 70's when Rolls-Royce faced extinction.

Happily, all ended well. And now it appears that some of that spirit has finally rubbed off on Guy Parsons, the man who did so much to save her.

The great Rolls-Royce liquidation came to an end last week when the Peat, Marwick, Mitchell partner, together with joint liquidators Ian Watt of Thomson McLintock and Thornton Baker's Maurice Withall, secured court approval for the distribution of the remaining £6¼ million. After a final shareholder's meeting in October, that, as they say, will be that.

Well, [illegible] conditions of the court is that I cannot be discharged,' explained Parsons who has been in charge of proceedings for the best part of the 11-year run. 'My only concern now is that I might want to retire before it is all over.'

The problem centres around possible claims against the underwriters over three aircrashes involving Rolls-Royce engines. In which case, said Parsons, 'I would be required to act as a post office and officially pass the claims on to them. But there are no more funds left now, so effectively it is ended.'

Looking back over 20 years in the business, Parsons admits that 'Rolls Royce was rather special'.

Perhaps most gratifying was his initial involvement with the motor car side under the original receiver, another Peats man, Rupert Nicholson. 'We decided to form a new concern,' said Parsons. 'It made profits from day one and was eventually launched onto the stock exchange. That's never happened before, I think.'

Guy's most prized possession was the Silver Lady Guy received from Rolls Royce to thank him for his involvement and success in helping them to achieve Independence of the Company with the inscription:

"From all your friends at Rolls Royce Motors Limited. In gratitude for helping to achieve Independence of the Company. May 1973":

78. Guy Parsons; Newspaper article regarding winding up of Rolls-Royce liquidation and picture of Guy's most treasured possession.

COPYRIGHT ACKNOWLEDGEMENTS

Images 55–56 copyright, The Sir Henry Royce Memorial Foundation, supplied with kind permission of Mrs. Brenda Wright.

Image 59–67 copyright *Derby Evening Telegraph*

Images 68–71 with kind permission of Kevin Miller.

Images 72-75, copyright receiver & joint liquidators with kind permission of the Rolls-Royce Heritage Trust.

Image 76, copyright Rolls-Royce, copy supplied by Charlie Hunt's family.

Images 77-78 supplied by and with kind permission of Philip Parsons.

CHAPTER 8

RECOLLECTIONS MEMORIES AND PERSPECTIVES ON THE ROLLS-ROYCE RECEIVERSHIP PERIOD BY FORMER ROLLS-ROYCE EMPLOYEES, TRADES UNION OFFICIALS AND OTHER KEY PEOPLE

THE AUTHOR, ALAN EDGE (Former Senior Contracts Officer, North and South America)

Memories of the Receivership as a Trainee

"I joined Rolls-Royce in August 1970 as a 15-year-old junior clerical trainee. Initially based at the Rolls-Royce Training School in Mickleover, I was one of an intake of about six trainees managed by a very well-respected training officer, a Miss Tunnicliffe, as our training leader. The training involved two monthly attachments to several departments throughout the two-year training period, after which you were expected to take up a permanent position.

My introduction to working life began at the Precision Casting Facility (PCF) on an attachment with the production engineering team tasked with an investigation into the use of 'bobbins' in the foundry wax room. The aim of this project was trying to reduce the number of bobbins used in the 'lost-wax process'. Not having an engineering background, it certainly 'lost' me initially but then, as always, help and assistance was never far away.

In November 1970, I moved on to work on a secondment within the Sinfin 'A' Site Services Team, where I remained until April 1971. By then everyone was starting to become aware of the problems and seriousness of the situation with the RB211 development/production programme. I spent my time working between the Site Services office, 'A' Site Receiving Stores, and 'C' Site ATF Stores.

At the start of February 1971, rumours were rife amongst employees as well as in the local press and national media that Rolls-Royce was in big financial trouble. So, as I walked home on the evening of February 3rd, 1971, looking forward to celebrating my sixteenth birthday, I wondered what was going to develop in the coming days and was my traineeship under threat so soon after starting my working life?

February 4th, 1971 dawned all too soon as I recovered from my previous night's celebrations; already there were widespread reports on the national radio of an engineering giant close to bankruptcy. My walk to work around 7.00am took me through the Moor Lane car park and past the Directors entrance at Moor Lane G Block, where, in the cold February morning, it became apparent that something major was happening; there were already people in groups, talking animatedly as I passed.

Crossing over Victory Road at the Midland Bank, I could see there were press and reporters and an outside TV van close to A Site factory gates, attempting to interview employees as they arrived for work. That day I was working in the Receiving Stores which began work at 7:30am. All the storekeepers were in deep discussions as I arrived and Ray Lee, the Stores Foreman, had brought in a transistor radio to monitor any developments. Sure enough, at 8:00am the BBC news confirmed our worst fears that Rolls-Royce had gone bankrupt! By then, work had already started with the unloading of two large delivery lorries, no doubt containing many vital supplies for the RB211 programme which was well underway. Stunned silence followed the announcement for several minutes followed then by Ray Lee saying to the delivery drivers; "Right, there are two choices here. Either we continue to unload your deliveries and along with it the risk that your company may get delayed payment or at worse never be paid for the goods or we can help you reload the lorries and leave the company premises immediately". They took the latter option and vacated the site, by which point the Rolls-Royce security men had sealed all the factory site gates,(as I now know this is documented in the receiver's memoirs and was on the instruction of a manager from Cooper Bros, accountants who were reviewing the financial situation of the company and advising the Government. This action had unfortunate consequences for vital deliveries). For the rest of the day there was no real work undertaken and many have said, for many days after that, as everyone was so stunned by the announcement. As I left work for the day, I decided to take a bus from the factory into Derby centre to see what was happening on the streets and town centre. However, it transpired the news was still all contained within the Sinfin sites and nothing much was happening in the town centre at that point.

In the weeks that followed, it became clear that all apprentices/trainees, etc. were to have their jobs protected. In the weeks after February 4th, I recall one particular task that I became involved in as part of the receiver, Rupert Nicholson's sale of saleable assets which lasted about a week on site. I remember accompanying a storeman and continuously travelling the length and breadth of the Sinfin A and C sites in a truck with a large trailer, visiting every department and secretary that was there. We were tasked with ensuring that every secretary returned to us all surplus stationery in their department and the instruction was to allow each employee only one pen, one pencil and notebook. You can imagine the reception we got from many secretaries reluctant to return anything and who tried desperately to hide their personal stationery stashes from us. I lost count of the full truckloads of stationery that we sequestered during those four days and failed to see what effect the sale of second-hand stationery would have on the huge financial crisis Rolls-Royce had found itself in.

Later on, in my career at Rolls-Royce, I worked in the commercial department and would manage major RB211 customers for the Lockheed L-1011-200 and -500 series aircraft including Middle Eastern airline customers Saudia of Saudi Arabia and Alia of Jordan."

SUE ALLEN Former Rolls-Royce Engineer – Further Recollections

"In February 1971, I was working in Turbine Engineering in B wing of Moor Lane. On the previous day we had heard rumours that 'something' was happening because managers were being summoned to secret meetings. Soon after arrival on Thursday 4th we found that the switchboard wouldn't allow any external calls; in 1971 only supervisors were allowed access to an external phone and all calls went via the exchange. I can't remember what time we heard the news of the bankruptcy that morning, but I do know that not much work was done for the rest of the day ... or the day after. Many of us were in denial, refusing to believe that our company would cease to exist.

I joined in the great march in London soon after February 4th to plead for the Tory government to nationalise us. Led by our Chief Negotiator of the ASTMS (Association of Scientific, Technical and Managerial Staff), we travelled to London aboard many trains and buses to march from St. Pancras to The Friends' Meeting House on Euston Road where we were addressed by several speakers including the late Anthony Wedgwood Benn who was the Labour MP for Bristol Southeast and so had a real interest in saving Rolls-Royce jobs.

We then marched down to Parliament where the police informed us that we couldn't display political placards within a mile of Westminster. I recall turning mine round where I had quickly written 'More pay for policemen' and they let me carry on! Whether or not we had any influence on the decision to nationalise us, we didn't know but we liked to think so. Redundancies were announced soon afterwards.

'First in, last out' was still very much part of the selection process but there was another strange feature; all married women were asked to leave. I had been married for seven months.

The news was given out quite brutally. In one area on Sinfin A, all staff were gathered in a conference room and names were read out publicly in alphabetical order along the lines of for example "A ... you can stay, B ... you're out, C ... you can stay but you'll move to another area", etc. It took a long time and must have been very tough on anyone whose name began with Y! All those made redundant had to leave by March 31st, 1971, in effect eight weeks after the crash.

It is difficult in the twenty-first century to imagine women losing their jobs because they were married although it was just a fact of life then. The view was that, after marriage we only had children and then resigned with no maternity rights.

Rolls-Royce was always an excellent employer and not unusual in the way that it managed women; discrimination was common practice in the UK until the Sex Discrimination Act 1975.

After the redundancy, I returned to the company in 1974, retired in 2004, worked for another 15 years as a contractor and am now involved with the Heritage Trust. I hope that the company viewed its support and investment in my education as an engineer as a profitable one."

JIM CLINTON – Spare Parts Pricing Department, Elton Road. Some Thoughts/Recollections on February 4th, 1971

"I had just become a departmental manager and we had to make some people redundant. Possibly it was still married women first (it certainly was in the 1960s) and I had to tell one that she was redundant, only to find that her husband had also just been made

redundant. Fortunately, someone in the department volunteered for redundancy to start a small business and therefore that woman could stay. I think that we got our required number for the department from volunteers.

On a lighter note, the milk float stopped coming on site (at Elton Road) as he was worried that the float might be seized by the receiver. Similarly, the chocolate machines had labels on stating their owners and that they did not belong to Rolls-Royce.

There was a rule that you took your furniture with you when changing job but the person I took over from was still sorting out an office. Somehow, I had got nice furniture, but it was in store and everything there was sold by the receiver, and I stayed with the old stuff.

It was lucky that we did not lose too many people as the RB211 was massively redesigned and every spare new part had to be priced before initial provisioning spares recommendations could be sent to the airlines."

ANGIE PAGE – Daughter of the Managing Director of Rolls-Royce Supplier, Mark Tyzack Ltd.

"My father, Dan Westmoreland, was Manager of a Rolls-Royce supplier, Mark Tyzack, in Sheffield, a small company specialising in precision engineering. The company was very proud of its long association with Rolls-Royce and, having been born in Derby, my father was particularly proud of the connection. I remember watching the Farnborough Airshow with him on television, with him saying, 'There's a bit of Mark Tyzack's!'. Dad was an electrical engineer and a very 'hands-on' manager. He was often found on the factory floor with his sleeves rolled up. As a consequence, he had a very close relationship with his workforce.

As teenagers we heard the news about the difficulties at Rolls-Royce on the radio but didn't really realise the seriousness of the situation. I remember men coming to the house in the evening, and Mum telling us to be very quiet as Dad had some serious talking to do with them. I later learned they were from the trade union and had come to talk about the work situation. Mark Tyzack had other contracts, but most of the work was for the new engine at Rolls-Royce. Without that work my father had no choice but to reduce the workforce. He came home and cried the day that he broke the news to them."

BARRY NEW Former Senior Vice President, Government Relations, Rolls-Royce, North America Inc.

"I was attending a meeting with Jack Brunning of BAC, Weybridge, on February 4th, 1971. Jack was well known for holding our 'feet-to-the-fire' on any BAC1-11 engine-related problem and that morning had been fairly typical of such meetings. He was called out while we were at lunch and was very quiet when he returned. He had been advised of the Rolls-Royce bankruptcy and was not sure how to break the news to us. He knew, however, that he had to say something, and he was extremely solicitous. One of his main concerns was whether BEA (British European Airways) would honour our tickets for our return journey to Glasgow. As it happened there was no problem. We later surmised that BEA owed Rolls-Royce a lot more money than the price of a few airline tickets!

In the days that followed we were called to a meeting with Don McLean who was the head of the Scottish factories at the time and were advised that we had to reduce numbers. We were also told those we retained were to be the strongest contributors, regardless of years of service. As a result, I found myself having to tell a member of my team, who had been with the company for 22 years, that he no longer had a job. This had even more impact as I had also been partly responsible for recruiting the team when we moved Civil Spey Engineering from Derby to Hamilton some three years earlier. Fortunately, it was not long before those who were laid off found good, and reportedly better paid, jobs in different fields–not surprising when you consider the company's ability to attract the best of the best.

From a personal perspective I, like many of my colleagues, wondered what the implications would be for me. Should I be looking for alternative employment? The options were very limited if one wanted to stay in the aircraft engine business. We were aware, however, that the RAF was offering direct entry at Squadron Leader rank for experienced engineers. At first, this seemed a promising prospect until I learned that an Engineering Squadron Leader had far less scope of responsibility than I had already enjoyed for some time. As a result, I decided not to pursue it further and, much later, I was given to understand that none of those who took the offer were ever promoted above their initial Squadron Leader rank.

In the months that followed we were instructed to clean out all the old material, without exception, from our filing systems. I was never sure whether the instruction was meant to be executed so completely but fortunately some people found ways to preserve some of the company history, even if it meant taking documents home."

BEV RIDGEWAY (Secretarial Trainee, 1971)

"I left Mackworth Secondary Modern School in July 1971 at the age of 14 and commenced work at Rolls-Royce as a secretarial trainee. My uncle, Ken Briggs, was an instructor at the Training Centre at Station Road, Mickleover, where the mechanical apprentices were being trained as well as the six secretarial trainees. He used to collect me from my home in Mackworth Estate on his way and it was great for me to arrive at the Training Centre in such a privileged fashion.

When the crash came in 1971, we were told the secretarial school would be closed and our future training was to be transferred to Wilmorton College. I was, on the recommendation of Miss Tunnicliffe, enrolled in the Advanced Secretarial Course. I later found out that all other general students who were not affiliated with Rolls-Royce had to have far more tertiary qualifications than I possessed. It was purely due to Miss Tunnicliffe's power that I was recommended to this course.

Whilst we six Rolls-Royce students were in full-time college education, it was great when the payroll officer attended the college on a Thursday to bring our brown pay packets. This was a welcome privilege over other general students not in employment. However, when the college broke up for holidays we were seconded to numerous and various sites within the Rolls-Royce Derby site for work experience. I can recall specifically working in an office in the Bassett Block in Moor Lane and having a tour of the computer block, seeing so many people in white coats because you were entering an area that had to be kept at a certain temperature. The computers seemed to be huge with big reels of what seemed like tapes. I am sure that compared to today's standards they were only holding a minuscule amount of information even though they were so big. I can remember how the internal mail worked with a daily postman delivering envelopes addressed to personnel on the various sites. The envelopes were reused many times with the previous address crossed out and the new one entered. I also worked at No. 2 works and was introduced to a telex machine. At that time, it seemed unbelievable to me how we could connect all over the world. I also remember how we had to get rid of the old stationery when the Rolls-Royce (1971) Limited company was formed.

Once college resumed, we would revert to college life again. As far as I was concerned, I don't think I would have got such extensive training at the Training College in Mickleover as I received at Wilmorton College. I can remember how everyone was queuing up at the Derbyshire Building Society in Derby, me and my family included, to withdraw our savings after the February 4th 'receivership' announcement. I also met Colin, my future husband of 44 years, at Wilmorton College. The Rolls-Royce crash of 1971 did in the end have a real positive effect on my life."

BILL MEHAN

My Bristol Memories of the Early Days of the RB211 Engine

"I am R. W. (Bill) Meehan and started my five-year apprenticeship (1956–1961) with Bristol Aeroplane Company (BAC) as a sheet metal worker, transferring to the Tool Room halfway through my time, ending my apprenticeship as a tool maker. Working mainly in the Brabazan hangar (aka Aircraft Assembly Hall or Concorde Hangar), these were very interesting times, and my work covered many various aircraft types such as the Bristol Britannia, Bristol Type 188, English Electric Lightning, Canberra, VC10, BAC1-11 and TSR 2, ending with the early days of Concorde. My first job as an apprentice in the Tool Room was making tooling to cheapen production of Lotus Elite car bodies being produced in glass reinforced plastic (GRP).

About 1966–67, I realised that the staff had much better privileges than those employed on the shop floor, so I transferred to 'weekly staff' status in the Jig and Tool Design office and was there until 1969. Wishing to get married, I realised that financially it would be easier to earn money as a contractor.

Plastics

Bristol Aeroplane Company had started a plastics department in the early 1950s headed up by Jack Rees (he had written several books and was a pioneer on the subject). I had been taken on by a Bristol company, Throsper Engineering, and was seconded to Bristol Aeroplane Company, Plastics Division, starting there a week before my wedding and honeymoon. I was told that on returning, I would be in a new factory with a new company name Rolls-Royce Composite Materials at Avonmouth, near Bristol.

The site

This was based at a modern-style glass-and-light structure head office facing Shire Hampton Way. Behind and at a right angle, was the production factory, not connected to head office except by pathway. From memory, the production area was divided into two, by a partition, with access doors. This area ran alongside and parallel to the M5 Motorway. The Planning and Design offices, including Jig and Tool, were located on a mezzanine floor, at the south end of the factory floor space. Three other buildings were within the boundary fence with car parking outside. First the laboratory and test facilities, employing only 10–20 technicians, were located to the west of the factory. Alongside those were the canteen and also the security and personnel department (now known as the HR department). The personnel department was headed up by Doug Stacey.

There were many composite products being manufactured on site and these seemed to increase in numbers daily. Some examples of products made include:

Artesian well pipes

Having been designed for the oil well business, these pipes were on a constantly running production line and could easily be made into a method of extracting water in arid climates by the simple cutting of fine slots. They were purchased by Pakistan and many developing countries. These pipes came in different diameters and were easily jointed without adhesives. They have even been used at great heights on the top of television repeater masts in this country.

Gun turrets for naval ships

This GRP was laid up as a skin, inner and outer with a balsa wood core of only modest thickness. They were able to deflect the shell and protect the gun.

Armour of various types

This included body armour (said to have saved President Reagan's life in the assassination attempt) and bulletproof clipboards. Plastic combat helmets were developed for defence requirements.

Motorcycle fairings

Made before the Rolls-Royce takeover were, however, still for sale.

Concrete pipes

Pipes for underground use were developed and tested, being found to be much stronger when reinforced with glass filaments during the concrete mixing.

Leaf springs for road vehicles

As carbon fibre GRP was being introduced, it became apparent that road vehicle springs made from this material would be much more resistant to the harmful effects of water, grit and other 'nasties' and would not be damaged so easily by them.

Drop tanks for aircraft made from phenol asbestos.

These were made for many aircraft and many other countries. They were manufactured in such a way that when disassembled the units would fit inside each other, taking up much less space for storage and transportation.

RB199 engine for the Tornado aircraft

Many acoustic panels were made for this engine.

Rolls-Royce and the RB211 Engine

As we were settling into our new design and planning office, we were asked by Rolls-Royce Derby to look at various components for the RB211 engine. The biggest component was the large fan containment ring with acoustic panels built in. This really was the leading edge of technology at the time, and it was unknown if blade containment and noise suppression could be achieved with this fairly light structure. Many other panels for the RB211 engine and nacelle were laid up with glass reinforced plastic (GRP) and honeycomb.

Large fan blade for the RB211

The method of laying up carbon weft along with tooling for these blades together with the heat presses all came from Derby, where they had initially manufactured and tested the blade. Our Bristol prototype blade seemed to satisfy everyone with the quality and finish being acceptable. However, after rigorous testing at Derby, it was found that the leading edge could not withstand the harsh conditions that were expected, and degradation of the leading edge (L/E) was unacceptable.

Soon blades with metal L/E were seen in many departments but whether they were produced at Bristol or Derby, I do not know. These were considered to be the answer to the degradation problem, but alas, after testing (I believe again at Derby), they were deemed not acceptable.

At this time, Rolls-Royce was in severe trouble, as was Lockheed in the States. Dan Haughton, the Chairman of Lockheed, visited Composite Materials at Avonmouth and called a mass meeting of every employee, both staff and works, together with contractors, spelling out to them how dire the situation was for our two companies and also the two countries. This meeting seemed to be the start of the decline of Rolls-Royce Composite Materials (RRCM), Avonmouth. Soon after, how long I cannot remember, most contractors had their contracts terminated and returned to their various companies. I returned to Throsper Engineering and after several weeks I was asked if I would like to return to RRCM, but as permanent staff. I gladly accepted this offer as contracts were becoming scarce.

The number of personnel remaining in the drawing and planning office was greatly reduced and my job had now become tool procurement, placing the order and progressing the manufacture. For this, our section divided the country into four quarters. I had the South-East (Bristol to London and down to Southampton), with many tools and fixtures being made and continually looking for firms with capacity and the skills required (not as easy as you would think). However, things were easing slightly as Boeing was now well into the 747 build, so spare capacity was becoming available.

All of this tooling was for engine nacelle panels and not the large fan blade. It soon became apparent that the drawing and planning of tooling manufacture was being returned to Derby along with the Jig and Tool drawing office and all associated personnel.

My personal view was that there always seemed to be slight friction between Derby people and Bristol employees, and we certainly had different ways of getting to the same end result. Two things during my happy but hectic few years with RRCM will remain with me. My immediate tooling bosses were the best management people that I have ever, in my working life, had the pleasure of working for, although life was very busy and deadlines very tight.

I have to say, although the RB211 engine was at a vulnerable stage at this time, it would seem the Ministry of Defence work was prioritised. The second thing, without wishing to make the differences between

Bristol and Derby people worse, was, it had to be said that the most important concern the 'Derby men' had was getting back to see the mid-week Derby County football match in the evening.

In conclusion, although the RB211 engine has turned out many versions in the subsequent years under the banner of Trent, it would seem one of the latest versions hit a stumbling block. As I am of the older generation – (I am allowed, I believe, to say), I hope this is not history repeating itself, as now the 'knife and fork engineers' have faded into oblivion. I feel the new designers have too much trust in their computers and little seat-of-their-pants experience.

Rolls-Royce Composite site at Avonmouth has now been demolished with nothing but the concrete slab it stood on remaining. Purely by chance, the new land speed contender car, 'Bloodhound', intending to reach 1000 mph, is being built within a hundred yards of the RRCM site, using a Rolls-Royce engine. We wish!"

DAVID READE

My Bristol and Rolls-Royce Receivership Experiences

"The year 1959 had seen the Armstrong Siddeley takeover of the Bristol Aeroplane Company (BAC) Engines Division, based at Patchway near Bristol (later to be known as North Avon in the 1970s). This was to supplement its other outputs from its Coventry sites at Ansty and Parkside.

The merger to be known as Bristol Siddeley Engines (BSE) had established an undercurrent of resentment from most of the workforce at Patchway after having been 'separated' from the BAC Aircraft Division sited up the road at Filton in Bristol. The Aircraft Division was later to become part of British Aerospace and ultimately a division of EADS (European Aeronautic Defence and Space, later, the Airbus Group). It would make the airframe and wing structures for the Concorde and the A300 Series, including the A380. As stated, the 1966 takeover had caused some long-running resentment to develop, relating to the loss of the other half of the Bristol Aircraft Company often referred to as 'family'. This was to persist for some years and continued into further takeovers.

My personal observation of this had started long before this period from my experiences with my father, brother, sister, uncles and cousins who had all established a long family tradition of working for

BSE. My father, who by then had nearly 30 years of employment, had expressed his concerns to me about joining BSE and had attempted to discourage me from seeking an apprenticeship with them in 1963.

February 10th, 1964, saw me commence my apprenticeship with BSE, (much to my father's disappointment) a career which was to ultimately last until 2001 on my retirement at age 54. I completed my apprenticeship in 1968 and was given three options: to retrain, take redundancy or take a test to join the technical staff. Following a successful test I moved onto the technical staff in the position of a planning engineer, based in the Planning Office of the Experimental Fabrication Department at Patchway. 1968 also saw my marriage at the age of 21 years, followed two years later by the birth of our first daughter. The personal events of 1968 were to set the picture for the dramatic company events that were to follow within the coming years at Bristol.

In 1966 we experienced yet another takeover of BSE by Rolls-Royce Ltd. based at Derby. Rolls-Royce had always been seen as our biggest competitor even during World War II. Within a few years of the takeover, through its campaign to sell the RB211-powered engine into the US market, Rolls-Royce made what was seen at the time as a sound commercial judgement in basing its 'sell price' for the RB211 engine on a $ exchange rate. This was to prove catastrophic during a period of currency fluctuations and subsequently helped push Rolls-Royce into receivership in 1971. This caused extreme concerns and fears for the 'Bristol family' workers for their jobs, homes and families and pushed the long-established resentment of takeovers further into the minds of the workers at Bristol. Backing from bankers or governments, as subsequently provided to UK banks after the later 2008 financial crisis, was not available back then in 1971.

I believe in the 1970s, non-government ownership of any company was not possible beyond a 40% limit of the company's overall shareholding.

A recovery was eventually made in the formation of Rolls-Royce (1971) Ltd. to ensure the continuance of the company by the government and to ensure the continuation of the contract with Lockheed for the RB211 engines for the Lockheed TriStar. The Rolls-Royce Motor Car Division was sold off to raise capital, but with the Rolls-Royce Aerospace Division retaining the marque only, and not the Car Division. To ensure continuity of supply to Lockheed in the interim, engines were initially supplied at the astonishingly low nominal price of £1 each. Profit was non-existent and this was perpetuated for a considerable further number of engines being supplied.

News of this reached the Bristol Engine Division (BED) and further exasperated the workers when BED was solely producing military engines and was actually making £100 million profit per year, but this was being used to support the failing Rolls-Royce Civil Engine Division at Derby. Some considerable time later, profit from the RB211 did slowly begin to trickle into the 'books' at Derby but only after having been supported by the profit-making Military Division at Bristol and having been subjected to the embarrassment of published pictures showing Lockheed TriStars sitting on airport aprons in the USA with huge concrete blocks attached to 747 wings to hold the aircraft down whilst awaiting late delivery of engines from Derby.

The 'Bristol family' has carried on successfully since 1971 with the groundbreaking efforts in the development of the Olympus 593 for Concorde; the RB199 for the Tornado; the EJ200 for the Eurofighter Typhoon; and the JSF (F-35) power plant."

(Author: By way of explanation see quotes of J. B. Perkins in Chapter 3 regarding engine price. See also point 11 of Chronology of Key Events in Chapter 6; point 16 in Chapter 7 and footnotes[16 &17] in Chapter 10).

PETER COLSTON

Key Extracts from his career and RB211 Experiences, as recalled by his daughters Susan Carlson and Joanne Watts (*in italics*) and from notes made by their father, together with extracts from the book '*Some Old Bemrosians of Distinction*' as contributed to by their late mother.

"Our father joined Rolls-Royce on a full-time basis in 1955 on graduation from Manchester University. Studying Commerce (probably Business Studies/Economics now), he had taken an extra year out because he suffered with polio during his time there. He was born in Alvaston, Derby, as were many of the workforce at that time. He had been a former pupil at Bemrose School, gaining a good reputation whilst there for his acting, basketball and scholastic achievements. Amongst other admirable qualities, he was described as being 'energetic, enthusiastic, principled and concerned for others".

He had been noticed at Rolls-Royce during a school holiday job having been turned down as a porter for British Rail. Rolls-Royce took notice of his progression through university.

There was a lot of prejudice around disabled people during that era. He was in a wheelchair for two years, which is why he did not do his National Service. He was told he was unemployable owing to the physical effects of polio, but he taught us that it may be a reason but never an excuse for not trying. He would later have his suits specially built up to hide his arm which had been affected. He was known for dressing smartly with bright linings to his suits, but it was actually to draw notice away from his infirmity.

After his degree he was keen to return to Derby to 'give something back to the town which had paid for his education'.

He made an early impact at Rolls-Royce, being selected as an executive assistant to the Managing Director, Sir David Huddie, between 1966 and 1968, and following a successful time in that position, he was made Divisional Output Control Manager in 1968. In our father's words:

'1968 was a very active time for Rolls-Royce as the company had secured a contract to supply engines for the RAF into the McDonnell Douglas Phantom aircraft. Deliveries were running 10 months late and at that time I was put in as Divisional Output Control Manager. In the British environment of the day, the delay in deliveries wouldn't have mattered particularly as airframes were notoriously late, but with an American organisation, this performance was unforgiveable. By developing new techniques and by integrating activities in 'the line', with functional control of production, we were able to bring the contract deficit from 10 months back to only three months. This meant producing 40 Spey Phantom engines a month. Still a significant record ...'

Our father was appointed RB211 Programme Manager in 1971, taking over the project at a critical time immediately after the receivership. His work was to ensure key objectives were met on time. He was assigned executives from all departments – design, development, experimental, manufacturing, production control and purchase – who were accountable and would all report into him. Each department and executive would be reviewed as to what was needed to achieve flight certification, engine type testing approval and delivery of engines to production aircraft.

'At that time, the RB211 (programme) accounted for over 60% of the total company activity in engineering, design, development, engine production and product support and therefore financial spend ...'

Our father would visit Sir David Huddie's home every weekend to discuss progress of the RB211 programme and keep him informed. The family lived in the Dower House at Edensor, on the Chatsworth Estate.

Susan: *I recall Dan Haughton, Chairman of Lockheed, visiting our house but I knew Archie Folden (Senior Executive at Lockheed and their 'trouble shooter'), much better and our two families remained friends until Archie died. Mother was a fantastic cook and Dad loved to entertain people at home, especially when they were travelling over weekends. I recall one occasion when Rolls-Royce was negotiating to supply industrial RB211 engines to pump oil for the Trans-Siberian pipeline. Negotiations were taking place at Duffield Bank House and had stalled. The Russian delegate was present, along with his KGB minder and an interpreter. The delegate joked asking, "Wouldn't your wife be wondering where you are?", so my father invited him to meet her. A little notice wouldn't have gone amiss as we children were ready for bed. It turned out that the Russian didn't actually need an interpreter, but the KGB man could only say "Hello" and "Goodbye". Sadly, Rolls-Royce didn't win that contract as it went to General Electric.*

Our father and mother, as well as I recall, Harry Malley, Head of the Experimental Department, who was quite a character, worked tirelessly to help people through charitable work. I don't believe many people knew they did a lot of fundraising between them for various charities.

From my father I learnt the great respect he had for the workforce as he would say 'They are the ones that make a company' *and he would know everything about them and their families and could talk to any one of them.*

After the financial crash, initial steps were taken within Rolls-Royce to develop an effective project/programme management system. My father worked very closely with the team from Lockheed, initially some 23 people, on-site in Derby to achieve this.

'Lockheed . . . were determined that Rolls-Royce would not default on contractual deliveries of engines ... We developed an excellent working relationship and rapidly the numbers of 'Lockheedians' were reduced as we were able to gain their confidence.

Apart from managing the (RB211) programme, I had to support the company's senior management in making regular reports to airlines, the British government, international banking communities, the supplier network and politicians. The lessons from working in such an open environment, with a wide variety of audiences, proved to be invaluable.'

Our father then became Material Director, Aero Engine Division Operations between 1975 and 1976. This involved oversight of production control systems and planning management, finance and purchasing. As a result of his efforts in introducing new methods and working processes, achieving immediate economies and laying down the foundations for the long-term evolution of the control over the Roll-Royce manufacturing operation, he received a commendation recorded in the board's minutes.

Contacts with other aircraft builders were given a high priority when he was offered the position of Commercial Director in 1975 through 1976 and he did a lot of what he termed 'bridge building'. It was during this time he made his first contact with Boeing. Apparently, he said it was a difficult period for sales, but planning was introduced to ensure Rolls-Royce understood customer needs and the chances of securing contracts from competitors. For a year, Rolls-Royce achieved 50% of the market share of engines with the RB211-524 engine, sold to Boeing for its 747 aircraft. Our father went on to hold further directorships within Rolls-Royce from Programme Director and Manufacturing Director, through to working at Rohr Industries (eventually becoming Vice President of Business Development) – a major component supplier to Rolls-Royce. He established an entirely autonomous team at Rohr in Chula Vista, California, in January 1982, working on the RB211-535 project, where he had full delegated responsibility for the subcontractors, achieving design specifications, testing criteria and ensuring production supplies were on time, in budget and at the required quality.

'Rolls-Royce, having secured the position (of supplying the RB211-535C) as launch engine in the Boeing 757 aircraft, had a significant programme to achieve. The task was made even more formidable when Pratt & Whitney launched an engine which claimed to have an 8% better fuel consumption than any engine Rolls-Royce could produce. Therefore Rolls-Royce, from a standing start, had to complete its original programme and introduce a new engine, the RB211-535E4, good enough to beat off the competition of the Pratt & Whitney PW2037 engine.'

The project was successful and left Rohr Industries as a free-standing supplier.

Our father loved the American way of doing business and their way of life. He sadly died in 1989 at 57 having undergone two bouts of heart bypass surgery, complications from the second surgery causing his death. His heart had been damaged by polio and by an earlier heart attack. He knew he would not live to be old and so made the most of every day. He lived the last five years of his life in America loving every minute of that time. He was buried in La Jolla, California, and over 600 people attended from all over the world to pay their respects, with floral tributes snaking around the block. We echo the comment in a book about former Bemrose pupils that 'his life was one of application of courage, resolve and integrity that enabled him to overcome the odds that would have crushed and overwhelmed many a lesser man'.

I don't think people realised how he and others worked tirelessly on the RB211 project to keep the company going immediately following the 1971 bankruptcy.

He very much enjoyed the last five years of his life in America but often reflected that his work for Rolls-Royce during the time after the financial crash was the most exciting period of his career."

CHARLES CUDDINGTON (Former President, Civil Aerospace)

"In 1971, I was a second-year student at Oxford University on a sandwich apprenticeship course of education and industrial experience, contracted to Rolls-Royce. On February 4th, as the momentous bankruptcy event was unfolding in Derby, I was completely unaware! Like most students, keeping up with the news was not a high priority. So, exactly when I became aware of the Rolls-Royce bankruptcy is now lost in the mists of time. I am sure someone told me, but what I do remember was that it was not until around May 1971 that I received a letter from the company setting out its proposals in the light of the bankruptcy.

I had signed a contract with Rolls-Royce Ltd. which obliged me to work one year with the company, then during the three years at university to work the first summer vacation with another suitable company, the second summer vacation back at Rolls-Royce and finally, after graduation, to spend a year at Rolls-Royce training in various departments before final placement in a job. In exchange for this commitment, Rolls-Royce would pay me a bursary of £100 a year whilst at university.

The letter basically said that as Rolls-Royce had gone bankrupt, it could no longer hold me to the contract. However, I could elect to continue on the sandwich course, receiving the £100 bursary for the final year, but with no guarantee of a job when I graduated. Of course, I would have to work the summer of 1971 at Rolls-Royce. The previous summer I had worked at a margarine factory owned by Unilever, which is a story in itself. It was a good experience but perhaps one of the impressions that shaped my future actions was that the issues around making margarine were far less interesting than those I had already been exposed to at Rolls-Royce.

So, maybe through student idleness or perhaps through some deeply subconscious belief in an interesting future of Rolls-Royce, I elected to stay the course. I found myself working for Dave Griffiths in Planning Inventory and Production Control (PIPC). We were located in the 'Hut' at Elton Road. I cannot remember too much about the attachment, certainly not the work! Impressions still linger though. Dave was an excellent chap to work for and he and the team worked hard and conscientiously, whilst also having some good laughs. I remember being told that the Hut was due for demolition. It was around not just for years but decades after that statement!

I finished my course at university and duly returned to the company in September 1972. I think it was probably at this point I realised that all the sandwich students in the year above me had been made redundant–well I think that was the case for Derby, but no doubt someone will correct me if I am wrong. It was a sobering moment. Others of my year turned up merely to resign–that did not compute! Of my contemporaries that jumped ship, I remember that one went to British Leyland (not a good long-term decision!); one went to W.D. & H.O. Wills, and one became a vicar! I needed no such divine intervention. Another colleague who elected to stay with Rolls-Royce thought he was being very smart. He asked if he could be transferred from Derby to Bristol. His logic was that as the RB211 was the root cause of the bankruptcy, there was no future in commercial engines in Derby, but there would always be a need for military products. His wish was granted, and he found himself working on the Olympus 593 – the Concorde engine. Oh, what irony! Needless to say, he left the company! The 1968 undergraduate apprentice intake had been the largest ever in Rolls-Royce's history. By the end of 1972 there were very few of us left."

MARGARET CUDDINGTON

"Unlike Charles, I was in the middle of the 'action' when the awful day dawned. I had just started working for a new director, who had moved to Derby from Leavesden, his title being Director of Government Contracts. This was in the days when you didn't get much choice about who you worked for, and I thought the work was going to be a bit dry and uninteresting after having worked in Sales and Marketing (as it was then). Our office was on the first floor of F Wing at Moor Lane.

I think the day Rolls-Royce crashed was the most eerie experience: no phones either in or out of the company and, of course, no mobiles in those days! The senior managers were going mad trying to find out what was happening. It was what they call today a 'lockdown'. As the day progressed the rumour mill started, and all sorts of reasons were floating around the corridors. I am sure the girls on the switchboard must have been given some idea of what was happening because they had to field all the incoming calls.

Later in the day all the directors were called into the boardroom to be told what was going to happen. I'm not sure when the receiver was appointed but my boss, because of his government contracts experience, was one of a team which was allocated to work with the receiver, together with me as his secretary."

DAVE PARFITT

"Toward the end of 1970, I was sent to work for Rolls-Royce based at Lockheed, Burbank, California, to replace Alan Parker who was the Engineering Representative.

In 1972, following Ken Spooner's heart attack and his return to Derby, I was made the Senior Representative at Burbank liaising with the Lockheed senior management. This lasted until 1974 when I returned to Derby as Assistant Programme Manager on the RB211. Bob Fryer replaced me, on a temporary basis until Bob Ward became available.

In my engineering role at Burbank, I worked with Howard Gray and his team, Bill Tooher, Fred Hendershot, and others, and with the Lockheed certification team. We had regular engineering meetings when Peter King, Rolls-Royce, senior project design engineer, came over and discussed the design aspects with Howard and his team. These were excellent meetings as King always agreed a course of action with Lockheed including who was responsible for each action. Amongst others, I attended meetings on the engine specification when Ian Cunningham, Contracts Manager, was a frequent visitor.

On the February 4th, 1971, I heard that Rolls-Royce had gone into receivership when I was driving to Burbank from my home in Santa Monica. When I got to the office the mood was rather sombre to say the least. I don't recall much of that day except the car hire company wanted its vehicle returned, which I declined. There was also the problem of company credit cards being unusable.

Ian Cunningham was the principle negotiator on the contracts side but unfortunately, he passed away a few years ago. Richard Turner was also involved on the commercial side. Mike Keen came over to run the flight test programme at Palmdale. The last I heard of him was that he was living in West Port on the American West Coast."

TED PAGE BY HIS SON IAN PAGE

1971 Experiences

"At the time of the Rolls-Royce financial collapse, my father, Ted Page, worked in the then Control Systems & Accessories Department (CS&A), where he was involved with the technical specifications for vendor-supplied accessories. Given the timing, this would have been on the RB211-22B and subsequent derivatives.

This role involved working with a range of vendors based in Western USA, and one such company I can remember is Garrett AiResearch, which was based in Phoenix, Arizona. I know that during the 1960s my father made regular trips to Phoenix, probably totalling in excess of thirty. One such trip took place in early 1971, and Dad and his colleagues were actually travelling back from the USA at the time of the Rolls-Royce bankruptcy announcement. They had taken the flight back to the UK successfully, but on arrival at London Heathrow found that the usual chauffeur car was not there to meet them.

This was, of course, because such supplier/service companies had no guarantee that they would be paid for any further Rolls-Royce work, nor indeed for any services recently provided. Based on this, such companies refused to provide any further services until a guaranteed payment mechanism was in place. Many suppliers at the time were left as creditors, chasing the receiver for their money. So, the only choice for my father and his travelling companions was to pool their personal resources, get to London St. Pancras, and from there a train back to Derby. Nobody would accept anything 'on account' to Rolls-Royce. Luckily, they managed to make the trip home, albeit somewhat delayed.

Obviously, once the initial crisis had been overcome, and Rolls-Royce was again solvent, the business needs dictated that liaison visits with accessory vendors must continue, and father engaged in this area of the business. His final role was on the commercial rather than technical side, until his retirement in 1984, after almost 50 years of service at Rolls-Royce".

DAVE TOMLINSON

"I joined Rolls-Royce in March 1968 and went straight into the Experimental Department as a skilled fitter. I remember my relief as I got my trade union green card to show I was accepted as skilled. I had a very unpleasant six months at the plastics facility on Alfreton Road where I was involved with some of the early RB211 plastic components. I thought at the time "These will never work". The rest of my time as a fitter was spent on Victory Road. I never worked on the experimental RB211 but worked instead on military projects like the Spey for the Phantom, the Tyne for the Belfast and the Adour for the Jaguar. In fact, I built the first non-reheated Adour engine destined for the Hawk aircraft still used by the Red Arrows.

The day the crash was announced I had gone to work as usual and, if I remember, it was at about eleven o'clock when there was a Tannoy message that the company had gone into liquidation. The effect stunned everyone. How could a prestigious company like Rolls-Royce go bust? Everybody stood around in little groups wondering and surmising what was going to happen next.

As the days rolled on, our department carried on working as near normal as we could with the threat of redundancy hanging over us, because at that time it was accepted practice to apply a 'last in, first out' policy. With less than three years' service, I was resigned to the fact that I would soon be unemployed. I started looking for jobs back in the motor trade as a motor vehicle technician as this was my previous trade. However, there didn't seem much enthusiasm within the motor trade for employing anyone new. It was as if the whole of Derby was waiting to see what was going to happen next.

Our department was very lucky as we were working on 'cost-plus contracts', whereby whatever the cost of the project, Rolls-Royce got that plus an agreed amount of profit. The result of this meant we kept working, including some limited overtime to get projects completed. The redundancy situation seemed to go away from my point of view. The company was asking for volunteers and although there was some compulsory redundancy elsewhere, it didn't really have any effect on our department or me. I only knew of two fitters who took the redundancy package, two brothers who were ex-motor mechanics who I believe saw it as an opportunity to get some cash. Both went to New Zealand. I didn't bank at the Derbyshire Building Society but watched the panic of people queuing to withdraw money on the television.

I stayed in Experimental until 1976, when I went on the staff as a technical author. In 1982, I moved to be a customer support engineer with Product Support. Then in 1986 I joined procurement in the Controls Department as a quality liaison engineer, moving through the promotion process where I achieved principal manufacturing project engineer status. I left that role in 1999 to become the Quality Systems Manager for Controls, where I remained until I took voluntary redundancy in April 2001. I have to comment that I fully enjoyed my time at Rolls-Royce and got to see a lot of the world whilst being paid!"

DAVID WHETTON (Former Executive Vice President Airlines – Rolls-Royce North America Inc., 43 Years' Service)

"I joined Rolls-Royce in Derby in 1963 as an engineering apprentice. Following an unfortunate 'accident' in the chemistry lab at Derby School, it had been suggested that I should forget about entering the sixth form and look for other opportunities. Rolls-Royce seemed like a convenient place to generate some temporary income while contemplating my future, but the attraction of the weekly pay packet overcame my desire to leave and do great things and I stayed with Rolls-Royce, ultimately taking a four-year sandwich course at Loughborough University, along with three other members of the apprentice programme.

All four of us returned to Derby in mid-1969 and gradually found our first real jobs. I was assigned to the Performance Department, but I had already decided that I was not a natural engineer. At a 'Meet the Management' evening for new graduates, I met Dennis Jackson, who had recently been appointed General Manager Sales. He told me that he was looking for an unofficial PA and offered me a position in Sales Support. I moved to B Site the following Monday, much to the displeasure of the Chief Performance Engineer and his HR department.

One of my tasks was taking the minutes of the weekly sales meeting which always included an update on the RB211 development programme. Ralph Robins (now Sir Ralph, of course) was the Deputy General Manager Sales at the time and he also gave periodic briefings on his efforts to keep the committed customers and Lockheed on board with the RB211. There was little good news and, by late 1970, it was clear that there would be major delays in certification and also significant performance shortfalls. This led to much discussion about contract delay damages, particularly through the Lockheed agreement.

Despite my familiarity with these issues and their seriousness, it was still a great surprise to me when, on February 4th, 1971, the company announced that it had filed for bankruptcy. I have few memories of that day and, surprisingly, do not recall how or where I heard the news. I do recollect the sense of shock throughout the office and later that evening on the television news. I also recall seeing people wandering aimlessly down Victory Road staring at the buildings, probably wondering, like me, what all of this meant for the future.

The day that I remember well was the one on which the redundancies were announced. My desk was at the very end of C Wing. Throughout the day, those who were being laid off (and there were many) were called to Sir Ralph's office and given the bad news. It became clear to me that a pattern was developing, in that the calls ran slowly along A and B Wings and then into C Wing. It was likely, therefore, that I would be one of the last to know my fate. Unsurprisingly, not a lot of work was being done and I spent a good deal of time talking to my three Loughborough colleagues who were watching a similar process taking place in their departments. By early afternoon all three of them had learned that they were being terminated, so I packed up personal belongings from my desk and waited for the call. By mid-afternoon there seemed to be a lull in the proceedings so I called Marie Topham, Sir Ralph's secretary, to ask how long I might have to wait. She told me that he had left for the day, but she had nothing to offer about my status. I arrived at the office the next day still thinking that I would be made redundant but was told that I was still employed. To this day I have no idea what criteria were used to determine who should stay and who was to be let go. We were led to believe that there was an element of 'last in, first out' but I had no longer tenure than my Loughborough friends or several others in Sales and Service who were let go. Some years later, I was shown a highly critical memo from the Chief Performance Engineer that had been placed in my employment file complaining about the way I had left his department, so the record there was also not in my favour. Maybe it was just luck, and I can only be grateful that I survived.

Despite the significant reduction in head count and talent, in the days and weeks following the redundancies the essential work seemed to get done, suggesting that there had previously been a lot of wasted effort (or maybe people just started working harder). I was assigned for a while to the office of the receiver and spent my time working through the archives looking for contracts and other documents required for the reorganisation. It was this experience that spurred me to join the contracts group a couple of years later and shaped the rest of my career."

DAVID DAYKIN

"Well, the development from the RB211-06 through to the RB211-22 was moving along slowly and in RB211 Development we were all aware that a crisis was looming. Hyfil fan blade rain erosion testing had resulted in blades looking like paint brushes. Leading edge strengthening helped, but the material was not considered acceptable. Going to the titanium blades added all the extra weight. Reverting from a wrapped aluminium fan casing with Hyfil blades to a 0.625inches thick steel containment section, steel fan casing for the titanium blades, together with strengthening of the front bearing housing were the main culprits for this additional weight. Of course, all the containment testing had to be repeated in the vacuum pit and the 'blade off 'test was only just kept within the confines of the casing! To compensate for the extra weight, the engine performance had to improve and forged high pressure (HP) turbine blades were on their temperature limit. We could not complete a full 150-hour-type test with forged blades. Cast blades, with much improved internal cooling, were needed but not yet available.

All the changes were causing financial difficulties with the project, which came to a climax on February 4th, 1971. The RB211 Chief Engineer stood on my desk in our Development Office and announced that the company had gone into receivership. We were all shocked, although we were aware of the difficulties both at Derby and Palmdale, with Lockheed, trying to satisfy future customer demands for more power and meeting the existing contractual requirements. We worked under the cloud of receivership until September 1971, when we were back to working overtime to try and meet the demands of the project.

In 1973, I moved over to the IP compressor section for the development of the -524 engine. This was a very interesting period and, just before I left the company for the first time in January 1978, we had the Federal Aviation Authority, North Western Region, carrying out Hot Re-slams on bodies of the Boeing 747 at 45,000 feet 0.85Mn(Mach number). This resulted in the intermediate pressure (IP) compressor, variable inlet guide vanes (VIGVs) operating outside of the P4/PF/angle tramlines. Initially this was overcome by slowing down the engine response rate on the re-acceleration and finally by using the feedback loop, i.e., using P4/P1 versus angle, as the control parameters, P1 being the pressure downstream of the IP compressor VIVGs."

DAVID SMITH (FORMER ROLLS-ROYCE DRAUGHTSMAN)

Receivership Recollections

"On February 4th, 1971, my recollection was that all work stopped and for the 30-plus minutes after the announcement, people in small groups discussed what this might mean. I worked in the Technical Block (Drawing Office) situated on Nightingale Road. The Tech Block provided the technical support for the production of the various components produced at the Main Works – blades, shafts, gears, discs, etc. At the time, I worked as draughtsman in Product Centre 03 which produced the large components. Harold Hinkle was the manager.

At around 9.30 (ish), an announcement came over the Tannoy system saying that Rolls-Royce had been put into receivership. I remember Harold Hinkle coming to the PC03's drawing section, which must have comprised approximately 30 plus drawing boards and 20 desks, with a very worried look on his face and speaking to one of the senior section leaders. I was told that he had said that he was expecting an even more severe announcement talking about closure, but this may not be true.

I was shocked but had almost expected something like this to happen. I think that I had worked on drawings for the same component for the best part of a year. I had drawn the component on five or six sheets of A1 plastic paper, only to have to rub it out on several occasions when I received memos giving updated alterations. My feelings were that the design process seemed to be in disarray.

At the time the Tech Block was in some ways 'stuck in the past'. There were still some of the older staff that could remember the days when its staff dealt with the technical aspects of the Merlin engine. They would reminisce about the days when senior members would come to work at 10 o'clock in the morning and stay until midnight trying to sort out some problem. They would also talk about the negative rake milling of the Merlin's cylinder heads in No. 1 Shop which took place 20 years earlier.

The career pathway was difficult to access and promotion not easy to obtain. Perhaps to help alleviate this, in the autumn – I think – of 1970, an initiative called Career Development was introduced. I remember most of the Tech Block staff going to a meeting in the Welfare Hall one afternoon to be told how this would work and how the problem of 'square pegs in round holes' would be reduced. We all sat listening when suddenly Arthur Gamble, a man who liaised between the Tech Block and the Works stood up and said, "While you lot are talking about this rubbish, we've not got the tools to do the job properly on the shop floor", and then he stormed out of the room. I remember thinking that Arthur was probably right. We all knew that things were not going well at the company.

One of the draughtsmen in the Tech Block–Denis Lawson (RIP)– went down to the demonstration lobby of MPs in London. We all chipped in to cover his lost wages on this day and made a banner. There was some discussion as to what should go onto the banner and one of the women in the drawing store suggested: 'Mr. Heath get your teeth into the problem' (Edward Heath had large teeth) but in the end I think it was 'People not profit'.

Later that summer, we also had a collection for the Upper Clyde Shipyard. We sent a letter with our cash that concluded with the words: "From one lame duck to another–'Quack'" (Ted Heath, Prime Minister had been quoted as saying: "Lame ducks go to the wall"). Whilst in no way left-wing, the Tech Block contained a group of more radical thinking people than many other areas. A member of the staff in the drawing store, for example, had a father who had been Oswald Mosley's bodyguard when he visited Derby. People would talk about this and other political views. One of Derby's most prominent 'communists'–a man called Potts–had worked in the Tech Block and there was still a legacy of building second fronts against the Nazis during World War II. In addition, the DATA trade union was to the left of most other unions and gave out a free monthly journal. Even if one did not read this, the headlines were to the 'left'. I remember one reading 'My advice is to agree with your employer and move on'. This caption was complemented by a policeman with a pair of handcuffs. This was a response to the Industrial Relations Act legislation then in force.

All, I think, were worried but some saw the redundancy lump sum as a bonus. Several people took this to pay off their mortgages. I remember one man saying that he would never get such an opportunity again. The redundancy process was called 'cascading' whatever that means. A sad case was that of one of my friends, Raymond Disney, who was made redundant and, if memory serves me correctly, had only just got married and bought a house in Chaddesden. Raymond was one of Derby's first Mods and a real character despite being only small in stature.

In hindsight, I think that the receivership could have been avoided but the whole company needed to be restructured. There was, for want of a better description, an RAF snobbery about the place. If you worked on a lathe in No. 1 Shop on piece work and doing fortnights about, shift work, for example, you really earned your money. This was in sharp contrast to many of those who worked on the staff. Such a restructuring and identification of essential work was probably impossible, as was the ingrained attitudes amongst many of the workforce.

I served an apprenticeship with Rolls-Royce and have to say that it was probably the best course of training in the world. On the very first day, we were told how Rolls-Royce was special, more than just a job. We were reminded that Rolls-Royce had saved Britain in World War II with its Merlin engine and told to all be silent for one minute and think about the important work that we were all about to undertake. In the Rolls-Royce Drawing School were posters with the letters of the alphabet, with each letter giving a message–e.g. A is for accuracy–Is your work accurate? P is for punctuality–Don't be late for work! etc. The atmosphere was so intense that on a couple of occasions, I had nosebleeds! All dimension lines had to be 0.007 inches thick and all outlines 0.012 inches thick.

This resulted in highly motivated and skilled people but often the 'nth' degree of these precise skills was not really required, and they became a goal in themselves. The actual links, for example, between the Tech Block and the shop floor were not good and, rather than concentrating on producing perfect drawings, improving the way in which the Tech Block supported the production of aero engine components would have been a better management objective.

The career pathways were not clear, and it was all too easy to get stuck in a rut. This led to frustration and often a cynical approach to any new idea. I suppose that times had changed, and the management had not adapted to these changes. There was, for example, little if any job evaluation in terms of identifying the crucial tasks and there were many highly skilled tradesmen on the shop floor whose talents were wasted. There was no real assessment of the value for money which a job gave. I remember in No. 1 Shop there were posters saying the component that you are working on cost £1,100, but at the same time, the men producing these were on piece work and trying to work as quickly as possible. It is true that by 1971, piecework had been phased out. But the above example is one of the muddled thinking. It was, I think, this muddled thinking that made the 1971 crash inevitable.

I was told that the 'mad miller' (milling machine operator) ran out into Nightingale Road shouting "Everyone is redundant" but I did not witness this event. There certainly was a 'mad miller'. And he worked in No. 2 Shop. Every time you walked past his machine, he would shout, "Ay you!". If you turned around to see what he wanted, he would shout, "Not you, you stupid b...", and if you ignored him, he would shout, "You ignorant b ...". He did it to me on several occasions."

GRAHAM WESTON (Former Rolls-Royce Fitter/Tester, Experimental Department)

"As a trade apprentice at the age of sixteen, I commenced my apprenticeship with Rolls-Royce on September 4th, 1967, at Ascot Drive.

February 4th, 1971 was to me and fellow apprentices very significant, as we were soon to become skilled tradesmen and, as such, we could find ourselves in the front line for redundancy. The day started quite normally until the dreaded message that Rolls-Royce had gone into receivership came over the loudspeaker system. After much discussion with a like-minded colleague, it was decided to apply for other employment in Derby in case we were to be the first sacrificial lambs. This task duly carried out, there was much debate as to who had secured the best promise of employment in the area.

We then noticed that the share price of Rolls-Royce had plummeted to a low of around 10 pence, and again after much discussion with fellow apprentices, it was decided that with most of the armed forces powered by Rolls-Royce engines, the government would have to stand in with a bailout. With this in mind it was obvious that the shares would then rise. As newcomers to share dealing, I and several lads went into town and procured Rolls-Royce shares at the, then, price of sixteen pence per share. As apprentices we didn't have lots of spare money, so we all bought as many as we could afford. Then came the letter that Rolls-Royce shares had been frozen, and we all thought that serves us all right for dealing in something that none of us knew anything about. But after a year or so, it was announced that for each share Rolls-Royce would pay £1. Happy days. This was the start of my lifelong love of share dealing. I also survived redundancy and carried on in my employment at this wonderful firm.

On February 4th, 1971, I was working for the experimental fitting shop on B Site, Department 3630, and later went to work on development test beds at Sinfin Department 3305, where military and civil engines were put through their paces, including the RT172 engine which was to be used by the Red Arrows. I remember several RAF Red Arrows team members visiting number 39 test bed to see their future engine running.

I am now retired but still have fond memories of my time at Rolls-Royce."

DAVE ROSSER

"The Shares Department at Elton Road was my first attachment when I started with the company in 1968.

Len Bradley will always stick in my mind for his strict rules, one being that if you were doing a share transfer you were not allowed to talk. It was frowned on sometimes that the office windows were open. We were at Sandown Road when the company crashed, working in the Spares Department and all the external phones went dead so we could not make external calls.

Eventually we were told the company had gone bankrupt. I will always remember that day and afterwards the government stepping in and saving the company. RB211-22B developments costs were blamed for the crash.

With the company changing its name to Rolls-Royce (1971) Ltd., before they went out, any letters or correspondence had to be stamped E. R. Nicholson, Receiver and Manager."

MAUREEN UNDERWOOD (Rolls-Royce Shares Department)

"Mr. Bradley was the manager of the department. There were several females who had all retired from Rolls-Royce and a Mr. Pendlebury, also retired from Rolls-Royce. I think they had paid out the first shares payment, and I was involved in the second, third and fourth. During the payouts they had to recruit several temps to cope with the extra workload.

I was a typist, and everyone was employed to try and trace all the shareholders to whom payments had to be made. Obviously, some had died, others had moved house, some had even moved abroad and sometimes wills were requested to make sure the payout went to the right person(s). One shareholder had his Rolls-Royce share certificate framed (the payout was less than it cost to have the certificate framed in the first place). He decided not to have the payout at all as shareholders had to surrender their share certificates to be paid.

A letter was sent out several weeks before the payout to warn shareholders to confirm their addresses. If the person had moved abroad, the letter would go by surface mail and take several weeks to arrive. This caused a lot of problems. If cheques were sent out and if a change of address came back, then we had to find the payout letter in the box to put the new address on it. This took a long time and resulted in a lot of painful backs. Later, when staff left and numbers were getting thinned down, I was found a job in in the secretarial pool, at Moor Lane as a loan secretary."

GURMIT SINGH (Rolls-Royce Shares Department)

"In 1971, I was working in the Rolls-Royce Shares Department at Elton Road. I was really enjoying this job. On a daily basis we had to register shares bought and sold and then had to balance money 'In and Out'. The Manager, Mr. Bradley, was really good. In those days if you were a manager you were seen as somebody. One manager usually had a staff of around 40 people. Staff working hours usually started around 8am and finished at 4pm.

At the time of the crash, the Ordinary Rolls-Royce share dealings were suspended. There were special '02 Workers' shares' which the employees could still sell but the Shares Department could only give out a limited amount of money each month. I was put in charge of 02 Workers' shares and operated a counter service where employees would queue up to sell their shares. I felt really bad to turn employees away after the month's limit was reached and had to deal with angry employees.

The priority was to pay out to Workers' shares, and I am not sure whether people got any money back on Ordinary shares. In the end, the Shares Department closed down and I had to find another job. I managed to get a job in the Standard Data Records Office (SDRO) where there was a manager, myself and six female clerk/typists. We were sending out material specifications to suppliers/vendors when requested by buyers."

IAN ANDERSON (Former Vice President Customer Business, Rolls-Royce Inc., USA)

"I was a commercial apprentice at the time of the bankruptcy, working in the Cost Data Control Department on a three-month attachment at the Elton Road site. A friend of mine rushed into the office on that fateful morning and said the company had filed for bankruptcy and that we should get our cars out of the parking lot because they were about to be impounded –the first of many rumours. I decided my bike was safe! Our supervisor called a meeting and tried to tell us the situation, but he broke down in tears. These were people with families, mortgages and their whole lives wrapped up in Rolls-Royce now facing a very uncertain future.

We apprentices were under the impression (again we weren't sure if it was true) that we were not at risk of redundancy because we were government funded. Naivety went a long way. I did remember one woman telling me it was unfair that a 17-year-old like me could stay employed while she and her husband could lose their jobs–this was possibly in response to some insensitive comments on my part.

My father and mother were both employed by Rolls-Royce, but I'm convinced I saw it as more exciting than worrying. My father was a turner on the shop floor and had been there since he was aged 14, so I think we decided he would be among the last to go. My mother, however, was a clerk in the Sales and Service Department and did receive a pink slip from the then Commercial Director, Ralph Robins. She spoke of them all sitting quietly in the office and waiting for a phone to ring as the 'culling' rolled through the office. People would just get up from their desk and return a few minutes later and sit down stunned.

I was interviewed by the TV cameras outside the gates of Sinfin A site one lunchtime and was asked my opinion. Again, showing my youthful ignorance, I said that I thought "They should put Brian Clough in charge". This was, of course, the glory days for the 'Rams' and Clough was very vocal in stating that a primary responsibility for the town football club was to create a distraction from the worries for so many people. He always considered himself a champion of the working class.

It was only as I matured and reflected on the realities of that time that I realised what a truly worrying and disturbing time it was for so many families. I went on to work for Rolls-Royce for 40 years and have immense pride in the company and all it has achieved. That dark day was a wake-up call that was needed, to turn it from a locally driven, rather narrow-minded British employer to the worldwide industrial power it is today. I and my family are proud to have been a small part of it."

LYNN MCDONALD (Payroll Department)

"In 1971, I was working in the Rolls-Royce wages office based in Mowbray Street. On the Friday before the news broke that Rolls-Royce was in trouble, I took the cheques for the weekly wages to the directors' suite at Moor Lane for signing. I then went with a senior

colleague to the bank in Victory Road (it was the Midland Bank at that time).

There, after presenting the cheque as normal, we were stunned to be told that there wasn't any money available to us. Once we were back in the office, we discussed the situation amongst ourselves, all of us completely shocked. Later we were told by senior management that the situation was not to be discussed with anyone else outside the office.

My father also worked at Rolls-Royce, and I still lived at home. I remember being so worried about everything that had occurred and being unable to speak to my father about it. I went to work as normal on the Monday and sometime during that day we had a delivery of money. I assume management had sorted something out over the weekend.

We had to work overtime to make up for time lost but no one minded, we were just happy to have some money to make up the wage packets.

Below are the Rolls-Royce sites/premises that we visited to pay out salaries. These are some of the sites around Derby I remember.

The Old Hall, Burton Road, Littleover. This was sold to Derbyshire Fire Service after the crash.

Laurie House, Colyear Street.

Station Road, Mickleover. This was where all the apprentices were trained. I believe this was purchased by Nestlé and is now a housing estate.

Slack Lane. This was the printing works. They did catalogues, etc. This was eventually bought by the Rupert Murdoch Group.

Babington Lane. Currently the location of Derby Post Office. There is a courtyard between the businesses and there were offices off the courtyard.

The last one I can remember was on Osmaston Road. Heading towards Nightingale Road from Derby, it was on the left-hand side not far from the foundry, which I believe is the Rolls-Royce Foreman's Club.

Rolls-Royce also owned Duffield Bank House at Duffield, which still exists.

To conclude, neither my father nor I were made redundant."

NICK TARR

Rolls-Royce Derby (memories from the 1970s after the crash)

"I joined Rolls-Royce in Derby as an undergraduate apprentice in September 1973. The great majority of the company's effort was focused on the RB211 which was entering service on the Lockheed L-1011 TriStar. In early 1975, one of my industrial placements from university was, at zero notice, changed to a small programme management team in Moor Lane supporting Jim Keir, the programme manager. Rolls-Royce was finding it difficult to maintain deliveries of RB211s to Lockheed and there were in-service problems that required engine removals and corrective action.

The programme management process had been set up to cut across the traditional departmental teams to improve the responsiveness of the organisation. There was a large photograph of a Lockheed TriStar in the office, sent by the Lockheed CEO (Dan Haughton) to Dennis Head who was the Managing Director of the Derby Engine Division. The TriStar had no engines and written across it, by hand, was the message that went something like this: "Dennis, I thought you would like to see an example of the problem that Rolls-Royce has created for us". New problems had arisen, hence my allocation to the programme team. This was before the more structured project management processes were fully deployed in the company. The disc problem was managed through daily meetings of all the people that had some input into the solution, engineering, metallurgy, production, product support, etc. There were often around 30 people in the room, and it all seemed very chaotic to me.

After graduation, I decided that I wanted a business rather than an engineering career. My first job in 1977 was in Purchasing HQ in Derby. Rolls-Royce was struggling to control costs during a time of high inflation (as high as 16.5% in 1976) and I was negotiating price rises with suppliers that had to be much lower than the prevailing inflation rates. Around 1980, the Purchasing organisation made its first attempt to introduce more automation with mainframe computers for use by all the purchasing staff. A lot of training was provided and the transition to the new system was remarkably smooth with a significant reduction in paperwork. This must have been a real productivity improvement.

From the late 1970s to the early 1980s, Rolls-Royce Derby was developing two new variants of the RB211, the -535 and the -524. Again, these were the prestige programmes that had considerable resources allocated to them. The RB211-535 was for the Boeing 757 and was the launch engine on this new project, a first for Rolls-Royce. The RB211-524 was for heavier versions of the TriStar and, for the first time, gave us access to the Boeing 747 market. This was the beginning of the 'Ralph Robins strategy' to widen the product range and gain a much more significant market share against GE and Pratt & Whitney.

All the time, there was constant pressure to reduce costs but, in retrospect, it wasn't very sophisticated. There was little emphasis on process engineering and suppliers were often frustrated when they proposed minor design changes that would reduce costs, but Rolls-Royce would not participate with any enthusiasm. This did begin to change by the early 1980s and I remember Salter (the bathroom scale company that also had an automotive division) trying to produce lock plates to the Rolls-Royce standard.

In 1982, I was offered a job in Military Contracts to work on the Adour programme for the US Navy's new trainer (based on the BAE Hawk). This was when I realised that, although Rolls-Royce's engineering was world-class and easily a match for the American industry, we were a long way behind on programme management, including cost analysis and reporting. It would take 10 years for Rolls-Royce to update its systems and change its thinking to meet the US government's requirements."

PHIL RUFFLES (Former Rolls-Royce Ltd., Director of Engineering and Technology)

"I was the Project Development Engineer on the RB211 responsible for performance and turbines. At 10.00am on February 4th, 1971, Chris Webber, Chief Development Engineer, called the whole department together in the main office in T2 building on Sinfin 'A' Site. He stood on a desk and read out the notice which said that the company had called in the receiver. We were all somewhat shocked but went back to our desks to continue working. I had just purchased a new car (an Austin 1300) and my wife was expecting our first child, so I did have some concerns. Also, being uncertain over whether my job was at risk, I did apply for a job outside Rolls-Royce. However, my main focus was to battle on to resolve the problems on my bit of the engine.

The previous evening, I had been supervising the testing of Engine 10011 which had demonstrated a step improvement in performance sufficient to enable the flight test programme to continue. Ernest Eltis, the Director of Engineering, communicated this to the board but it was too late to change the course of events. Nevertheless, it was a critical factor in determining whether the programme should continue at all.

In the days that followed there was a clamp down on expenditure including reducing the number of development engines and allocating hardware budgets to all project development engineers. Cost control prior to this had been poor. Within no more than two weeks, a new programme had been agreed with the receiver and so the focus was on fixing a number of major problems, completing the flight test programme and achieving certification.

In parallel with this, the specification was renegotiated with Lockheed, the thrust requirement for initial engines was lowered (RB211-22C) and the Entry in Service date was delayed by six months. (In fact, we were only four months late to the original schedule.) The pace of activity increased as we realised it was a case of saving the programme and the company. Sir Stanley Hooker (who came out of retirement to become Technical Director) said in his book, 'Not Much of an Engineer', "Thus, between February and the autumn of 1971 the Derby works achieved a miracle in totally transforming the RB211 programme. The modifications needed extended to every part of the engine, but they were carried through with a speed and a spirit reminiscent of the Battle of Britain".

Everyone was completely committed and just got on with the job in hand. There was a big redundancy programme on a 'last in, first out' basis. Several of the young development team were notified that they were to be made redundant, but in most cases, they stayed as there were sufficient volunteers from older people. Nevertheless, the loss of talent and failure to attract the best engineers during the 1970s was apparent for many years after.

Hindsight is a marvellous thing. There were certainly lessons to be learned, the most important of which was the need to invest in technology ahead of product development."

ROBERT SEARLE BARNES
Memories of the Rolls-Royce Receivership, 1971

"In 1971, I was the Administration Manager for Rex Nicholson, Production Director, and a main board member. My job was to manage the personnel and administration functions across all the production factories of the aero- engine division including catering, security and transport at Derby, but not Hillington or East Kilbride.

On February 4th, 1971, I was called to the Divisional Company Headquarters at Moor Lane, given a sealed document and instructed to set up a meeting at 11.00am of all those who reported directly to Nicholson, including work centre managers, and read a prepared statement to them. (Rex Nicholson was, as a board member, away in London). The statement shocked everybody, there was a stunned silence, and the meeting broke up without any discussion. We all knew that the RB211 engine was in trouble and that the company needed to be bailed out, but no one I think really believed that Prime Minister Ted Heath would let the company go to the wall.

At this time Rolls-Royce had a very unusual management structure. It would not be accurate to use the word delegation, but de facto power rested with the individual directors of engineering, finance, production and commercial. Both Company Headquarters and Divisional Headquarters were small and provided a framework for the business rather than managed it. This again changed overnight as the Official Receiver, Rupert Nicholson, took control.

My next significant memory is being instructed to make more than 4,000 production employees redundant across England and Scotland. (Plus ça change, plus c'est la même chose). In this, Scotland did its own thing, but I managed what may have been the first ever computerised redundancy, at least in the UK, each work centre had its own Personnel & Administration (P&A) manager and we gathered together for marathon sessions sorting out who must stay and who must go. The unions naturally wanted a 'last in, first out' (LIFO) policy to be adopted. But they realised, like us, that to follow this strictly would be to destroy the business. Our most recent recruits included large numbers of specialists in computers, then a new technology, and virtually all would have disappeared if we had applied LIFO strictly. Therefore, we had to sort out the population into more than 120 job occupation groups and apply these, subject only to the need to keep the company functioning.

It was a nightmare job, not helped in my case by the new possibility that I myself might actually be on my way, so we were told that the P&A function would be addressed last after the completion of redundancies for those affected. One of my P&A managers at the foundry left soon after for the Health Service. From Edinburgh, he recalls that he had been in the graduate trainee recruitment drive in 1970 and that he had dreaded meeting any of the intake when they were made redundant so soon after he had promised them a good future with Rolls-Royce. From Northern Ireland, another P&A manager, who later managed the Dundonald factory in Northern Ireland, recalls that from there they just watched as bystanders. They were making parts for Derby Product Centre 01 (wheels, discs and shafts) and were totally unaffected! As for myself, I ended up as a beneficiary being promoted to do what I had been doing three managerial jobs before, combining divisional salaries administration, job evaluation and the huge function of training, including responsibility for the Mickleover site.

I think I should end by saying how very well the receiver, Rupert Nicholson, managed the company. I cannot give any detail but do recall that he seemed to know exactly what needed to be done and, of course, he had all the authority he needed, not being bound to seek board consensus. Decisions came quickly and I recall feeling how much better the company was being managed than before. One particular humorous story recalls working throughout the night to complete the redundancy list on time and in the early hours being visited by divisional director, Trevor Salt, who asked if there was anything he could get for us. In response I said, "A fish and chip supper please", which was duly delivered within the hour."

TONY LINTOTT (Former Head of Year 2000. Procurement) A personal view of 'The Crash' of Rolls-Royce

... "I turn now to the redundancies which have rightly concerned many Hon. Members. On the latest estimates which we have of the scale of redundancies, some 4,300 Rolls workers are likely to be dismissed between now and 26th March. In the main, these are engaged on work other than the RB211. The main areas affected are Derby 2,000, Glasgow 1,700, Barnoldswick 240, and various Midlands plants 400. What I think is necessary to emphasise is that these redundancies are directed at eliminating overstaffing and are an essential part of the reconstruction of the company on a sounder footing. The figure of 4,300 was announced by the receiver on 8th March, and these decisions are for him in consultation with Rolls-Royce (1971) Limited." *The Lord President of the Council and Leader of the House of Commons (Mr. W. Whitelaw).* HC Deb, (11 March 1971), Rolls-Royce Limited. [*Hansard*] *(Vol 813, cc 668-669).*

"Had I known this at the time my fears for my family and friends would have been much less, however my story begins a few years earlier.

I was working in a Purchasing Office in South Block, Elton Road, when I was summoned to the office of the Manager of Derby Machining, Ron Green. All slightly odd, as he wasn't my boss and I'd never had anything to do with him. He explained that as Rolls-Royce had, relatively recently, taken control of Bristol Siddeley there was a need to bring this group into the Rolls-Royce fold and way of working. Rolls-Royce would contain three entities: Aero Engine Division, Bristol Engine Division and Small Engine Division. One of the Bristol factories was at Sunderland in the Northeast where operations commenced in March 1951 for the manufacture of components. It also happened to be in an area of the country which regularly received development grants from UK HMG.

Ron Green further explained that this factory was a Rolls-Royce-nominated disc and ring facility, but they had no production purchasing ability, just local consumables. He went on to say that I was to go to Sunderland, on a permanent basis it seemed to me, to create and run a purchasing department, using the existing staff, which would have the capability of procuring all items required to support the factory (actually a group of four units), whatever these items may be. I was, to put it mildly, shocked. Why had this offer come from a manufacturing man and not my department head? Why would I want to go so far away from home? Why me? The questions raced through my head as he continued to tell me about my future, and no options. His final statement was that I was to be paid £1,800 per year, with annual reviews, under a three-year contract.

When I managed to get a word in, I asked as many questions as he would allow. The most urgent one was, of course, about the pay offer. This he had insisted was non-negotiable. I recalled a few facts I had in my memory and respectfully stated that £1,800 per year was far below the pay grade for this type of job. I was met with a stony face and reminded that I was, "too young to get the full rate". My reply, "Well you want me to do the job, so why can't you pay me appropriately?", was not well received and I was dismissed from his presence and told to await further instructions. I was somewhat bemused and annoyed at the same time. Eventually, the whole thing settled down. My head straightened itself sufficiently for me to accept that I would go to Sunderland and turn it into a career move. I also came to the conclusion that my time in a local band would have to end as getting to gigs would be impossible. This was another difficult decision as it brought in a few bob on top of my wages as well as being immensely enjoyable.

So it was that I ended up in my new office in the Rolls-Royce Sunderland factory in Pallion, in late 1969. My office was just to the right of the entrance, on the ground floor, to receive suppliers quickly and easily, etc. and to keep them out of the factory unless absolutely necessary. Time passed quite quickly, the job was immensely interesting, and I was appreciated by Ken Raine. He was appointed as Facility Manager on the same day as me, so we had an affinity. The people were brilliant, a great bunch, provided you could understand the lingo. I learned it all from Scott Dobson books, 'Larn Yerself Geordie' and the folks in my digs on Roker seafront. I also managed to get married to my wife who came from the Derby area, and she moved up north near to where her forefathers came from. It wasn't all bad.

Interestingly I soon found out I was only the second 'Rolls-Royce man' to appear on site, the first being Ken Hunt, the new Personnel Manager. It was fascinating to see how the locals responded to these strangers from down south who, no doubt, had come to do them out of a job, sooner or later. How wrong they were, as shown by the new incarnation of Sunderland at Washington today. I was subsequently followed by another Ken from Derby being Ken Young, to run production and a further Ken who came to run the drawing office for a while. As a Tony I was clearly the odd name out, but I did fit in, unlike some other 'transplants' over the years. Mike Moseley, who worked with me in Derby years later and in recent times Chief Operating Officer–Supply Chain, would know exactly how it felt when he went up north! 1970 moved on into 1971 and there was little sign of what was to come in a few weeks time. The vagaries of aerospace I'd got used to in my previous nine years at Rolls-Royce and the problems with the RB211 seemed a long way away, 164 miles to be precise. I could do nothing but do a good job and support the factory and my team.

At 10.00am on February 4th, 1971, all was to change.

As a member of the senior team, I was summoned to the upstairs dining room where all eleven of us in management would fit in round the table. We had no boardroom. Everyone (well almost) wondered why the gathering had been called at short notice. Ken Raine read from a prepared briefing document and pronounced the company in receivership. The receiver, Mr. Rupert Nicholson, had nominated our Chief Accountant (a Bristol legacy title) Matty Robinson to be his representative on site and no spend would be incurred or committed without his signed approval. For me, this meant all purchase orders would have to be justified and counter-signed by Robinson. Stunned silence followed, then murmuring, then intense discussion about our future, Sunderland's future and would Rolls-Royce disappear forever?

The next few days were quite surreal. At work the new receiver processes for keeping the factory going developed into understandings and became the norm. There were more meetings and interfaces with those directly involved with the receiver and I wore a track up and down the stairs to Matty Robinson's office. I also, somewhat coincidentally, had developed a very basic programme to run on the accounts department advanced calculating machine (an early computer I suppose) and from it I was able to provide reports of spend and commitment to show the receiver we were in control and doing what was required in the receivership process. Eventually and in a relatively short time – a matter of weeks if I remember correctly – Matty Robinson told me I did not have to bring all orders to him for countersignature; Sunderland was being treated differently, from my perspective.

A further sign that things at work were not going to change much was the lack of rumour or fact regarding local redundancies. As the weeks went by it became clear that people would be leaving Rolls-Royce, the company, but not Rolls-Royce, Sunderland, the close-knit, factory community in a deprived area of the country. And they never did. At home in our little miner's cottage, we'd bought with wedding gift money (incredibly cheap at £500), my wife and I sat each night to watch the news on our rented colour TV. We watched in amazement (and horror) at the queues forming outside Derby Building Society offices when people went to extract their money in case there was a run on the banks. On weekend visits to Derby, the gloom and doom was almost tangible. Then, as we raced back up the A1 (I had an MGB courtesy of band earnings) the depression lifted and we went back into the swing of North-East life, trips to Holy Island, and boozy nights in the Roker Hotel and so on. It seemed we were insulated from the demise of our proud company, but we could see there was a huge problem to overcome.

Throughout the months that followed, HMG and lots of others racked their brains over the receivership. As we all know, there was a resolution, and the company was 'saved' by nationalisation owing to its strategic importance. It has survived and prospered ever since, with a few hiccups on the way as the aero engine business always has. As 1971 progressed, the 'normality' of my time in Sunderland rolled on until one trip to Derby in late autumn 1971 mapped out my future and a return 'home'.

The crash shook the company hard, and change was imperative, if it was to survive. New post-crash procurement groupings were developed, and this is where the next stage of my career began. I brought all the production purchasing I'd controlled in Sunderland back with me to Derby and it was appropriately distributed within the new organisation. Perhaps some semblance of the 1960's organisation had been restored too, but in sad and difficult circumstances. However, it did include the Bristol Engine Division and Small Engine Division elements Ron Green had mentioned and was the model for the next few years, until further changes came along, of course.

Returning to my opening statement, it should now be clear that had I seen this during the early days of the crash, the stress levels would have been much lower. You see there was no mention or hint of Sunderland being affected. As it was, the stress wasn't too bad and much, much less than that of the poor souls who had their feet to the fire of redundancy.

There but for the grace of God.

Though I left Sunderland in late 1971, to this day I still keep in touch with friends up there. My landlady in Roker died in early 2018, well into her 90s. I still miss being in the North-East and I conclude that Ron Green actually did me a favour with his kick in the pants to 'go up north'."

Tony was one of my former managers in the 1970s. Sadly he died in March 2022. I'm grateful that I was able to document his recollections.

TONY HOLME

"At the time of the crash, I was working on Adour reheat systems in Fuel Systems Development based at Sinfin, Derby. My memories are of all the phones suddenly being cut off around 9.00am and no one knowing what was happening until later in the day when we received a briefing from management. There were some older colleagues who were very upset as they held significant savings in Rolls-Royce shares – seems familiar at present. I was one of several recent graduates and the cut-off for redundancy on a 'last in, first out' basis was immediately below me. One of my colleagues who was made redundant got a job immediately with Rolls-Royce & Associates on Raynesway and a £500 severance award, so he was quite happy.

Another colleague was stuck in France at Turbomeca, and no one could contact him or find out how he would survive, as all funds were frozen. He survived and returned a few weeks later. I attended several mass meetings organised by the unions and heard Phillip Whitehead speak. He did a good job representing our interests with the government.

I left the company a few months later in the summer of 1971 as I could see no immediate future but returned in 1977."

BILL SEAMAN (Former Chief of Compressor Transmission Engineering)

"I worked in the Adour Development Department in the T2 Block on Sinfin A Site and when I arrived at work on the morning of February 4th, 1971, everything seemed to be as normal (we were a bit remote from RB211 Development which was located in the 'Seven Storey' building). Nevertheless, as we walked through the Experimental Fitting Shop to check on the Adour engine build, we would pass the inspection tables.

On February 4th, these tables were the same as they had been for months, laden with the wreckage of the latest RB211 test failure. We thought, again, "This can't go on – something has to give". At about mid-morning, someone in the office said he couldn't get an outside line – "Must be the phone". Other phones were tried, and none worked. Shortly after that we were called together and told by our Project Engineer (Alec Collins) that the company had gone into receivership and there were no communications with the outside world. Obviously, there was much discussion and we talked to our RB211 colleagues, but they were as much in the dark as we were.

Despite the uncertainty and confusion, we carried on as near to normal as possible–there were still engines to be built and tested. Going home that night was chaotic with the media outside the gates trying to interview us but we couldn't tell them any more than they already knew. After February 4th, we carried on as normal, but it soon became obvious from both the media and Rolls-Royce management that there would be significant job losses, assuming the company kept going at all.

I can't remember the timescale, but it can only have been after a few weeks that we learnt how many jobs were going from each department, including Adour Development. When the fateful day arrived when people would learn whether they'd lost their jobs or not, we were all sitting at our desks (as though by being present it would somehow improve our chances). The Chief Development Engineer for the Adour (Alan Cotton) came into the main office and touched someone on the shoulder and asked him to go into his office. We all knew what was happening and every time Alan came into the office, everyone shrank into their seats, hoping he wouldn't touch them on the shoulder.

There were about 24 people who worked in the Adour office, and I think we lost about six, but I can't be sure. I was lucky and survived for another 31 wonderful years at Rolls-Royce, but I'll never forget the apprehension, bordering on terror, hoping I wouldn't get the touch on the shoulder!"

JOHN BEDDOES (Son of the Provost of Derby Cathedral, 1971)

"My father, the Very Rev. Ronald Alfred Beddoes, was Provost of Derby Cathedral at the time of the Rolls-Royce receivership. In 1971, Phillip Whitehead, MP for Derby North, and others had formed a working party to represent Rolls-Royce's case to central government and my father was co-opted onto a working group, along with the, then, Derby Mayor, Councillor Miss Eleanor Grimwood-Taylor. My father might have been the chair of this group.

The group consisted of a cross-party representation of local politicians and local councils and organisations who worked to put a rescue plan together. They sent a delegation to London to request a bailout plan for Rolls-Royce. I think my father went to see Iain Macleod, MP, who was Chancellor at the time, meeting him at the Treasury in Whitehall.

At some point around this time as part of its Harvest Festival celebrations, the Cathedral had a Rolls-Royce engine sited at the back. I do remember a lot of talk and conversations about Rolls-Royce and many phone calls. I recall that Roy Jenkins, MP, the Chancellor before Iain Macleod, came to stay at our house overnight and I am not sure if that may have had something to do with the receivership, but I think he came to stay in 1974 when he was Home Secretary as my mum was complaining to him about a proposed change in the clocks from British summer time.

Soon after the bankruptcy many people from Rolls-Royce still had to be redeployed and many became teachers or went to work at the local British Rail Works. When I was teaching at Nightingale Junior School in 1978, I got someone in to talk about careers in the rail industry and we were shown a film of the development of the 'Advanced Passenger Train' (APT). We were told that its development had been influenced by the engineers from Rolls-Royce who had transferred there to work."

STEVE SHEPHERD

"I joined Rolls-Royce in January 1970 straight from school as an undergraduate 'thick sandwich' (five years) apprentice. There were over 100 of us at the Mickleover Training Centre doing the basic machine shop and drawing office training before being placed into a couple of 'experience building' departments on A Site Experimental. We then went off to College in September 1970 where we received £100/year sponsorship from Rolls-Royce. I believe that the government funded such training.

The first contact we had after the bankruptcy was a letter from Rolls-Royce to say that our apprenticeship would continue with its £100 per year sponsorship until September 1973 when the company's position might have become clearer, although people were released from the commitment of a five-year apprenticeship at this point. There was, therefore, no need to get concerned about the situation but I knew that many of the others from the same intake severed their ties with the company even at this stage.

Ahead of the expected summer summons back to Derby we received another letter that told us to use the summer to get a job overseas. I managed to get a job for six weeks in Los Angeles with a company that made nose-wheel jacks for 747s and followed this with six weeks hitchhiking around the States. It wasn't until the summer of 1972 that I returned to the company for some further placements, by which time things had stabilised.

On return after graduation in September 1973, there were probably only around 10% remaining from our initial intake of apprentices. I managed to get a placement in the Turbine Office in Moor Lane and stayed there for seven years before branching out around the company where I had a very varied set of jobs over 42 years. I certainly never regretted keeping my options open and maintaining my apprenticeship with Rolls-Royce."

February 4th, 1971 – A 'Churchillian' Speech

Extract taken from 'Bygones ' *The Derby Evening Telegraph*, February 24th, 2011, supplied by Peter Craig, former Rolls-Royce employee.

... "It was a rousing and inspirational speech by one of the Rolls-Royce managers which Peter Craig of Derby still recalls from those dark days. Pete was an assistant foreman in the Experimental Engine Build and Rigging area. Harry Malley, who he thinks was the site manager, gave a Churchillian-type speech about the problems the firm faced and how it was not going to be beaten by them. He said, 'The speech was inspirational. At that time, we worked days and nights. After the speech the fitters, etc. said they would work three shifts for no extra payment to help overcome the problems. This they did for some considerable time during which the performance of the RB211 engine improved to an acceptable standard. This helped resolve some of the problems. It must be remembered that those RB211 engines are now the bedrock of today's aircraft engines and Rolls-Royce's present status as a world leader in its field.

People like Sir David Huddie, the head designers, chief engineers and all their staff should never be forgotten for the foresight they had in designing a three-shaft jet engine."

THE ROLLS-ROYCE TRADES UNION OFFICIALS & THEIR PERSPECTIVES ON THE RECEIVERSHIP PERIOD

"People sometimes knock us at Derby and tell us that we are company men. That is what we are and it is our company and our interests. Unions should not see themselves simply as opposition parties to management."

CHARLIE HUNT, *Amalgamated Union of Engineering Workers, Works Convener, Rolls-Royce (1965–1981). Former Chair, 1971, Rolls-Royce/Trades Unions (Save the RB211 Campaign).*

"When Charlie Hunt joined Rolls-Royce in 1951, he was determined about one thing. He wasn't going to become involved in union affairs again! He had been a shop steward at the Derby Locomotive Works prior to World War II. How little did he know then that it was going to be just the opposite. On retiring after 29 years dedicated service to Rolls-Royce, Charlie had established a reputation as one of the best-known grassroots trade unionists in Britain, respected enough to impress even Margaret Thatcher. It was reported that when Charlie met the, then, Prime Minister at a 10, Downing Street reception in the autumn of 1979 she said, "I've heard so much about you, I carry your words with me" and she produced from her handbag a newspaper cutting about him on industrial relations at Rolls-Royce.

Charlie was a man with family connections to Rolls-Royce going back to World War I and later he followed in their footsteps after firstly rejoining British Railways in 1945 and shortly moving on to Rolls-Royce. Most famously, Charlie had applied to join the Royal Navy in 1939 but was firstly refused on the grounds that he was in a 'reserved occupation'. He re-applied and finally convinced them stating his occupation as a 'pram fitter'.

Whilst serving on HMS Belfast (the ship that helped sink the German battleship Scharnhorst), he was also involved in supporting the D-Day and Normandy landings. When the war was over, he eventually rejoined Rolls-Royce intending to stay well clear of union work. However, it was not long before he took on the role of a shop steward in 1959, eventually becoming the Rolls-Royce Works Convener from 1965 to 1981. He was responsible for all the manual workers at Rolls-Royce's Derby, Mountsorrell and Castle Donington sites with a workforce then of around 11,000 people.

Charlie will be best remembered by many for chairing the trade unions' campaign to save Rolls-Royce after the 1971 crash. During this period the joint works and staff trade unions worked around the clock in the battle to save jobs. Rather than accept their fate, the Rolls-Royce trade unions were determined that the company would be saved. Plans were laid for lobbying local and national politicians; petitions were printed and launched in the city and surrounding towns and villages.

Charlie was revered by many and there was unanimous agreement that he was a very shrewd operator. He didn't have an aggressive negotiating style and wanted to work with the company to come up with a solution rather than get into a dispute. He was known to be very measured and thoughtful. He had a great ability to assess the situation and do the right thing. You just trusted his word and he had the unusual skill that he could negotiate without being confrontational. Later in his life he received the recognition which he truly deserved and was awarded the British Empire Medal for services to trade unionism in 1973. Four years later he received the Queen's Silver Jubilee Medal."

With acknowledgements to Alistair Hunt; The *Derby Evening Telegraph*; Eric Lancashire, MBE, and Mick Walker, former chief union negotiator & Mick Lomax (Former Unite [ASTMS] Trade Union Chief Negotiator)

ERIC LANCASHIRE, MBE (Former Chairman of Rolls-Royce TASS Trades Union Technical, Administrative and Supervisory Staff) By permission of his daughter, Gwyn Lancashire

"On February 4th, 1971, the Rolls-Royce management had requested that all staff and works unions attend a meeting at 10am in the board room at Nightingale Road. The room was full of union representatives from the Derby, Hucknall, Ilkeston, and Mountsorrel sites, including people like Charlie Hunt and Mick Walker.

Some had heard rumours, but no one spoke as we waited for 10am. Via a statement to the unions, the Managing Director announced that Rolls-Royce was bankrupt and was being placed into the hands of the receiver. Everyone was devastated and limitations were placed on all external communications from the company.

A special train was booked, and buses were organised to take people to lobby Parliament in London. I recall standing outside Derby Railway Station from about 5am with pockets bulging with money because the transport, train and buses, would not depart until they were paid, and I had to pay them.

On arrival it was a procession through the streets to Parliament with press and television coverage. The public openly supported us and cheered us on our way to the Houses of Parliament. When we arrived at St. Pancras Station, officials were met by a police superintendent who said, "Welcome to London gentlemen. My officers will escort you and may I wish you good luck in your campaign". I haven't heard of many demonstrations getting that sort of reception since.

One memory that stands out was when Charlie and I were invited to Lambeth Palace to meet the Archbishop of Canterbury. We were afforded every courtesy and encouragement. The meeting lasted several hours, and we had all quite forgotten about the time and felt the need to contact our wives and family to let them know that we were going to be extremely late returning to Derby. The Archbishop said, "Don't worry, I can contact any person in Christendom within an hour". Messages were sent and duly received by our families from church ministers in that time frame."

MICK WALKER (Former Rolls-Royce ASTMS [now UNITE] Union Chief Negotiator)

"I will always remember the day the receivership was announced. People had expected redundancies, maybe on quite a large scale, but not bankruptcy. When I left the PCF (Precision Casting Facility) at Sinfin where my office was based, to go to the meeting at Nightingale Road there was a low mist with a red sun coming out of it–it was like an omen.

At the gates, a bloke in a car stopped me and said, "We've gone bust". I said, "What?" and he repeated, "We've gone bust. I just heard it on the radio". I arrived at the meeting where there were 30 or 40 trade unionists from all over Rolls-Royce. I told them and no one believed me. Geoff Fawn, Rolls-Royce General Manager, and Dennis Hurd from Personnel arrived to break the news. I can remember one of them holding a packet of cigarettes and his hands were shaking so much he dropped the packet and sent cigarettes flying everywhere.

At the PCF, the resident engineer made an announcement over the public address system that the canteen had run out of chips. A wag next to me said, "Things are really getting serious now". First of all, though Rolls-Royce had to be saved; that was a priority. Then we had to persuade people that the RB211 project was important to the future of Rolls-Royce if it was to survive.

Redundancies were announced across the whole company, and this was a pretty messy exercise. There was no 'last in, first out' and no asking for volunteers. Then they decided to ask for volunteers, so people who had been told they were to go then got asked to stay because someone else had volunteered to leave. There was a right of appeal, and it could be heart rending, watching a man's life with the company being scrutinised. I had to represent lots of members and men would burst into tears when told they have lost their appeal.

Looking back, it was a case of everyone being absolutely stunned. The last thing we expected was for Rolls-Royce to go bankrupt and it had a tremendous impact on the town. Eventually the company was saved but there were still 2,000 job losses."

MICK LOMAX (Former Unite [ASTMS] Trade Union Chief Negotiator)

"I joined Rolls-Royce as a trade apprentice in May 1964 and left halfway through my apprenticeship to go to NOTSA Engineering, based in Aston-on-Trent, rejoining Rolls-Royce one year later to continue and finish my apprenticeship in November 1969. I left Rolls-Royce at Christmas 1969 and emigrated to South Africa in January 1970, working in South Africa until December 22nd, returning to the United Kingdom on Christmas Eve 1970. I rejoined Rolls-Royce on January 24th, 1971, just less than two weeks prior to the crash (not my fault).

I was working on the dayshift at the Nightingale Road, Main Works (No 1 Shop) as a tool grinder. Thursday February 4th was a strange day as I remember it. There were rumours that something big was about to happen, but no one could have prepared us for what did happen. When it was announced on the works Tannoy system at 10am that Rolls-Royce had gone bankrupt, I remember our foreman, Fred Tapping, coming out of his office and telling us to stop working. Like many others I didn't know what to do or what the future would hold for us.

Friday, February 5th, 1971, was another strange day with no work being done. I remember lots of management meetings happening, but nothing further was said to us on the shop floor.

In 1971, the Rolls-Royce board room was also at the Nightingale Road Main Works factory, but no one saw or heard from any of the company's senior management team.

The following couple of weeks were a little hazy. Work on the shop floor had started again but we were still not getting much information from management. The government had stated that there would be no bailout for Rolls-Royce. This was not the view of the unions which believed Rolls-Royce was worth saving and wanted to do something about it. Bucket collections had started and there were many trade unions mass meetings both in the yards and the Welfare Hall at Nightingale Road.

I had rejoined the union on January 25th, something nearly everyone did when starting work at Rolls-Royce and I was amazed at how quickly the Unions organised things.

The main speakers at the Derby site meetings were Charlie Hunt, the works convener, Archie Wynne, the joint shop stewards, chairman, Mick Walker, the chief negotiator for the ASTMS union, and Eric Lancashire, also a senior staff representative for the TASS union. These mass meetings were my first real involvement with the trade union movement and I do feel that it was listening to those men that made me want to be like them and represent the workers' voice (which I subsequently did over a long career in the Trade Union movement) The union meetings were joined by local church leaders, local councillors and local Members of Parliament, Phillip Whitehead, who was the main spokesperson and Walter Johnson. Buses and trains were organised and paid for by the Union from the bucket collections and local funds to lobby Parliament.

We continued to work as normally as possible following the February 4th announcement. This was mainly on the advice of the union as we wanted to show the government the true worth of supporting Rolls-Royce.

This was a very difficult time for my own family; my father, brother and two uncles also worked at Rolls-Royce Derby. Unfortunately, my brother and one of my uncles were made redundant and this was very upsetting but in those days the redundancy rules were very different from today's. At that time, the selection criteria for redundancy were; married women first, dilutee[15] *next, then semi-skilled and finally skilled workers on the 'last in, first out' (LIFO) basis. These criteria made me safe, despite only being there less than two weeks when the announcements were made.*

I have no doubt that had it not been for the Rolls-Royce trades unions at all the United Kingdom plants there would be no Rolls-Royce aero engines today."

15 The term "dilutee" was first used by the Air Ministry in October 1939, when the Temporary Relaxation of Existing Customs Agreement was signed. It denotes an industrial employee who has not served an apprenticeship or had equivalent engineering experience but is nevertheless employed on skilled work.

CHAPTER 9

TRUTHS, MYTHS AND CONSPIRACY SURROUNDING THE ROLLS-ROYCE LTD. RECEIVERSHIP

PRICE PAID

It would not be until around July 1973 that a settlement was reached covering the price to be paid by Rolls-Royce (1971) Ltd for the assets of the old Rolls-Royce Ltd under a heads of agreement dated 17th March 1971.

This was debated in the House of Commons on Wednesday 27 June 1973.

Mr. (now Lord) Tebbit MP asked the Secretary of State for Trade and Industry if he could make a statement on the negotiations between the receiver and Rolls-Royce (1971) Ltd. on the price to be paid by the company for the assets they acquired from the old company.

Mr. (now Lord) Michael Heseltine MP responded as follows.

> "Rolls-Royce (1971) Limited and the receiver and joint liquidators of Rolls-Royce Limited are today announcing that a settlement has been reached covering the price to be paid for the gas turbine business acquired by Rolls-Royce (1971) Limited under the heads of agreement dated 17th March 1971. The settlement, to which the Government are a party, incorporates the resolution of a large number of other claims and ancillary matters outstanding between the receiver and joint liquidators on the one hand and Rolls-Royce (1971) Limited and the Government on the other. The settlement provides for the payment to the receiver of an overall consideration of £87½ million, of which £30 million has already been paid on account of assets and undertakings transferred. Interest will be payable on the balance, and the sums due reduced by the sterling equivalent of $21 million in respect of advance payments on the RB211 will be paid to the receiver by Rolls-Royce (1971) Limited on or before 23rd July . . . .
>
> HC Deb (27th June 1973), Rolls-Royce Limited [*Hansard*] (*Vol 858, cc 343-5w*)

The proposal was put to the independent expert appointed by the Government and also a Lord Justice concerning certain legal matters. Both agreed the conclusion reached was "reasonable and fair to both sides.".

THE CONTRACT THAT NEARLY NEVER WAS!

Lockheed evaluation committee's engine choice.

RECOLLECTION OF ROBERT PAUL (*Former Vice President Materials, Lockheed Aircraft Programmes)*

"I am now in my late eighties, so it takes some time to recall such old history. Working on the Lockheed L-1011 was far more interesting than other programmes because of all the problems encountered over its production life and you could be creative on how our contracts were made up on commercial programmes, compared with government programmes. I loved what I did in my working career at Lockheed.

The Lockheed Chairman at the time of the Rolls-Royce engine selection for the TriStar was Dan Haughton and Carl Kotchen was President. When the Lockheed management met to present their collective conclusion for the engine selection for the L-1011 aircraft, they advised the Chairman that they had picked General Electric (GE), over Pratt & Whitney and Rolls-Royce. The Chairman asked why and was told there were questions over Rolls-Royce's financial ability to develop a new engine, with old manufacturing facilities and a limited history of commercial aircraft support. Dan Haughton said he had had several aircraft projects with GE engines over many years and they had failed Lockheed a number of times and were very hard to deal with. He overruled the selection committee's engine choice and said, 'We will pick Rolls-Royce with a new high-technology engine with a forecast of better fuel consumption and a lighter engine weight.' He added, 'We can link sales to British Airways in a large number to help launch the L-1011 and England can help us with other sales with countries they have links with.'

The Lockheed engine evaluation team was stunned! The selection committee was made up of Purchasing/Legal/Finance/Manufacturing/Engineering/Program Office/Marketing/Product Support and Quality, along with Kelly Johnson, the head of the Skunk Works. The Chairman relied on his selection committee's engineering reputation for such a technical selection and yet on this occasion he did not follow its recommendation."

FACT FINDING MISSION – Secor Browne

US President Richard Milhous Nixon asked the chief of the American Aeronautics Registration Board, Secor Browne to visit Rolls-Royce on a fact-finding mission. It was reported by the receiver that he had met Secor Browne, in what appears to have been a chance meeting, sharing a 'leisurely breakfast' on the 15th February 1978, at Duffield Bank House where they were both staying and reported that the talk was 'most reassuring' and fitted in with what Dan Haughton had said to him previously as preferring, if possible, the RB211 engine even if it cost more that the contract price despite being able to adapt to the GE engine if necessary. The receiver was left with the impression that there was 'an appreciable leeway in price and that the fuel consumption and noise were all favourable to the RB211'.

The GE engine was different in size and shape from the Rolls-Royce engine, as well as having other differences and would have required a redesign of the L-1011 to accept the GE engine. Lockheed would say this would add to cost and time, estimated at $100 million plus, with GE stating it would cost $52 million. Lockheed's re-evaluation of the RB211-22B, P&W JT9D and GE CF6-80 engines and having positive discussions with the Airlines, all decided it was cheaper, quicker and better to proceed with the RB211 than with an alternative. There was also a potential for sales of the L-1011 in countries with links to the UK, with attendant trade benefits to the US, a British engine in an American airframe and potential American sales within Europe.

FLIGHT International 20th May 1971 commented on this and quoted the chairman of Eastern airlines, Mr. Floyd D. Hall, having stated on May 6th 1971, of "the 'splendid progress' of the flight test programme and all other testing of the Rolls-Royce RB211 had again demonstrated the technical excellence of the aircraft".

US GOVERNMENT LOAN GUARANTEE: In the months following the receivership of Rolls-Royce Limited, there was mounting uncertainty concerning the continuation of the RB211 contract with Lockheed and indeed the future of Lockheed Corporation and its L-1011 TriStar aircraft programme. Lockheed were only able to continue with the Rolls-Royce engine following support from the US Government in the form of a loan Guarantee passed by congress on September 9th, 1971.

US CONGRESS VOTE

Robert Paul again recalls the following: *"During Lockheed's efforts to secure a US Government Loan Guarantee, I was calling all the chairmen of our suppliers in the US and other countries to lobby their members of Congress to vote for this Loan Guarantee. We used postal codes to match our supplier locations to Members of Congress. Only one company tried to undercut our support because they gained more if Boeing sales increased because of our failure. I spent 36 hours in the office (one night and day) making calls and tallying up who supported us and reporting results to the Corporation. We won the vote in Congress by* **ONE VOTE!**".

Some years later Roy Anderson, Chairman of Lockheed (1977-1985), referring to how close the vote in Congress was, told Peter Pugh that Dan Haughton had said to the Lockheed lawyer, 'Bring out the bankruptcy file'. In referring to the competition from MDC he thought, in his view, Rolls-Royce had 'busted their ass', and had produced 'a fine, long-lived engine'. Ibid *Pugh P. (2001) (pg.232).*

THERE WAS NO US GOVERNMENT SUBSIDY TO THE ROLLS-ROYCE ENGINE.

A price increase was agreed between Rolls-Royce, the Airlines & Lockheed. The proposed US Government guarantee of bank loans would be strictly for Lockheed, enabling it to meet the peak financing requirements for development and production of the L-1011 in the United States, which was beyond what Lockheed could borrow through normal commercial channels.

GE LATE CANVASSING TO PRESIDENT NIXON

Whilst Lockheed was actively evaluating alternative engines for the L-1011, given the uncertainty about the future of the RB211 or indeed Rolls-Royce's ability to supply a viable engine; General Electric made a late attempt in May 1971 in canvassing directly to President Nixon for a change of engine selection for the TriStar, suggesting as a replacement the GE CF6-80 engine, should Rolls-Royce or the RB211 project completely collapse.

A letter dated May 3rd, 1971, from the GE Chairman Fred Borch to US President Nixon, the contents of which have been extracted and published previously in *FLIGHT International, 27 May 1971* highlighted to the President the benefits of an 'All American' aircraft and drew attention to the Presidential request to Congress for guarantees of up $250 million for bank loans for Lockheed, deeming it as promoting an unfair advantage at the expense of US Industry.

However, as we now know, this lobbying was to be to no avail as Rolls-Royce and the RB211 survived and continued supplying engines for the contract for which it was, and for which it remained, the sole engine supplier. The RB211 engine would greatly contribute to noise reduction at airports and in surrounding areas.

'A CRISIS OF CONFIDENCE': THE DERBYSHIRE BUILDING SOCIETY

'In God we trust, the rest pay cash!'

(On a plinth in the office of Ian Cunningham, Rolls-Royce Contract Manager)

Within days of the bankruptcy of Rolls-Royce Ltd., rumours were circulating that the Derbyshire Building Society (DBS) was believed to hold hundreds of thousands of Rolls-Royce shares and was about to go into insolvency itself. Widespread panic developed amongst depositors and long queues formed in Irongate, Derby, with investors trying to withdraw their money. The situation became a classic run on a financial institution that lasted for three days at its height with many depositors returning to bring back what they had withdrawn.

"Roger Hollick, Marketing Officer for the DBS said, "What happened to the Derbyshire following the Rolls-Royce crash was heartbreaking, unnecessary and history making.

Rumour, fear, ignorance and some adverse media coverage fuelled public anxiety and some people thought 'if it could happen to Rolls-Royce it could happen to The Derbyshire'. . . We went up and down the queues persuading people to go home as there was nothing wrong. We had many successes. . . some people were embarrassed and said, "It's not for us, it's for someone else", when we talked to them in the queue.

People were taking away cheques and some just kept them at home until it was all over and then brought them back to us! I met lots of people who took money from us on demand and went and paid it into banks. When it was all over, they went to withdraw it to bring it back, only to find that they had to give notice to get the money. . . . At the very worst we would have been taken over by a very big society. . . . Some had lost money in Rolls-Royce, some had lost their jobs, . . . it became like a war that we were determined not to lose. There was no way that Derby's other great institution was going to succumb. . .

A crisis of confidence built on rumour is very hard to overcome".

***The Derby Evening Telegraph Special Anniversary Flashback Supplement,* February 4th, 1991.**

In Fact the DBS was barred by law from investing in industrial shares such as Rolls-Royce. It should also be said that only 1,500 out of the 27,000 mortgages portfolio held by the DBS were actually held by Rolls-Royce employees. The other security that it had as a member of the Building Society Association (BSA) was that if anything was wrong the other BSA members would have stepped in to protect savers and investors.

Following this the DBS went from strength to strength and managed to remain independent for some period before being taken over by the Nationwide Building Society in December 2008.

RB211-22 DELAY ESTIMATES: Despite previous delay estimates of between six months to a year in flight engines being delivered the engines were delivered to Lockheed in four months. Sir David Huddie told Peter Pugh in 1999, "I am not blaming anyone else . . . We had promised a bit more than we could perform. Things were not deplorable. Flight engines were four months late. Four months is four months, but it is not deplorable. The deplorable thing was the cost . . . The accountants never got cash-flow into our heads". Ibid *Pugh P. (2001) (pg. 156).*

In fact, the proto-type RB211 engines were delivered early in the receivership, in the second week of March 1971. Therefore, there was very little delay between Lockheed wanting the engines and Rolls–Royce delivering them.

Ibid *Nicholson ER. (1993) (pg. 53).*

During the renegotiation of the contract, the chairman of Trans World Airlines (TWA) Charles C. Tillinghast Jnr., had indicated to Dan Haughton he was not troubled by delay apparently even up to a year, given TWA's own financial difficulties, and he was more concerned to have the TriStar programme continue than what engine would actually power it. Ibid *Pugh P. (2001) (pgs. 230-231)*

THE GOVERNMENT WHITE PAPER CMND 4860 & DAMAGES:

Cmnd 4860 on the RB211–presented to Parliament in January 1972, referred to the damages which Lockheed and the airlines could claim. In fact, Dan Haughton repeatedly and publicly said that they, Lockheed "didn't want damages, they wanted engines".

BANKRUPTCY – PRE-CONCEIVED The events surrounding the receivership of Rolls-Royce, caused several conspiracy theories to develop as to why the company may have collapsed, all of which have been found to be groundless and unsubstantiated in the passage of time. One of these was that bankruptcy was deliberate but unnecessary.

Mr. John Stonehouse Labour MP had raised this point two years after the bankruptcy in the House of Commons on the 13th of February 1973 during a debate on Rolls- Royce.

"In the House over two years ago the then Minister of Aviation Supply made a statement of the utmost gravity – namely, that Rolls-Royce Ltd. was proposing to make itself bankrupt. That was because Rolls-Royce had been advised that it was unable to meet the expected liabilities under the RB211 engine contract which had been entered into with the Lockheed Corporation. In the statement, the Minister said:

> 'As the Board of Rolls-Royce have stated, the loss of resources already committed to the project combined with the losses which will arise on termination are on such a scale that they are likely to exceed the net tangible assets of the company. In the light of this situation the board have decided that they have no alternative but to ask the trustees of the debenture holders to appoint a Receiver and Manager.' [*Official Report, 4th February 1971; Vol 810, c1922*].

It has become quite clear since that date that the board of Rolls-Royce was ill-advised to go into bankruptcy, because it would have been possible to renegotiate the contract with the Lockheed Corporation without the extreme step of bankrupting almost the, if not the, leading engineering company in the United Kingdom. The effects of that bankruptcy have been felt by British exports in all corners of the globe. It is indeed regrettable that it took place. The bankruptcy undoubtedly took place because of the pressure that the

> Government put on Rolls-Royce and the advice that was given by the Government to the directors of Rolls-Royce at that time".

In rebutting this during the debate, Mr. Cranley Onslow, Parliamentary Under Secretary of State for Aerospace, thought Mr. Stonehouse's speech 'rather odd'. He went on to say:

> "He brought little new evidence to his cause this evening. He made an allegation about the Government having applied extreme pressure to force bankruptcy on the old Rolls-Royce company. I did not hear much evidence adduced for this. If he has some no doubt, he will not hesitate to give it to us. In the whole of his speech, I did not hear a single reference to the White Paper (*Cmnd. 4860)* published just over a year ago, which he must have read. If the Hon. Member did read it, I fear that he may have forgotten some of it, because he will find set out there a clear and succinct account of some of the matters which seem to worry him. I direct his attention, in particular, to paragraphs 18 onwards of that White Paper and suggest that he looks at what is written there".
>
> HC Deb 13th February 1973, Rolls-Royce Limited. *[Hansard] (Vol. 850, cc 122-42).*

In his book, Ibid (*Rolls-Royce Limited*: *Recollections of the Receiver and Manager) (pg. 52 & pg. 54)*, Rupert Nicholson stated that during his period as receiver he had come across the unfounded allegations of a preconceived plan to bankrupt Rolls-Royce Limited, as outlined below.

> *"The suggestion that the government deliberately bankrupted Rolls-Royce; this was not within my knowledge. It is vital to understand that at the beginning of February 1971, Rolls Royce was hopelessly bankrupt; creditors were being held off and they doubted their ability to pay the next week's wages. .*
>
> *. . . "My appointment as receiver might possibly have been, therefore, to carry into effect a plan conceived in the UK government to nationalise the company with Ian Morrow as their prospective managing director. The receiver would be obliged to exercise his power to repudiate the RB211 contract as onerous, handover the worthwhile assets to the government after reorganising them with a receiver's special power, (thus leaving) the receiver to pick up the debts and pay what he could to the creditors and to the numerous claims for damages, particularly the potential huge claims from Lockheed and the American airlines. Perhaps the receiver might*

have been expected to avoid these latter chores by handing over to a liquidator and taking his discharge.

. . . In fact, it never struck me as a plot until journalists (and others) put it to me much later; indeed, it was not true so far as I was concerned. The fact we conferred with Rolls-Royce (1971) Limited was quite natural in the circumstances and assisted me in securing the best price for the business. It is however of interest that sometime later, while Anthony Wedgwood Benn was at the Department of Trade and Industry, I was invited to a small gathering at his department which proved to be a study group of civil servants, bankers and other secondees, exploring receiverships as a cheap means of nationalisation. I was shocked and explained the terrible liquidation losses and damages entailed falling on innocent creditors.

It is a fact, however, that a Cooper's insolvency partner told me by telephone on 3rd February 1971, that the receivership should not be so difficult as it appeared; preparations had been made. Also, Lord Carrington might well say that he was willing to buy the aero engine business at a fair price, which must surely be reasonable, i.e., discounted, because the establishment regarded Rolls-Royce as a 'failed company' and the RB211 engine as a probable disaster. The government, out of their concern for the national interest, were going to try to salvage the best of the business and I, as the receiver, might have been thought to be in a poor position to argue.".

ORDINARY SHARES, WORKERS SHARES & SUB-CONTRACTOR PAYMENTS, RECEIVER'S EXPLANATION.

By the end of 1982 each of the 80,000 shareholders received 64.5p for each one, ordinary share held.

The issue of the substantial worker shareholders and the army of sub-contractors was also raised in the aforementioned debate both by MP's Mr. John Stonehouse and Mr. Phillip Whitehead.

To avoid substantial delays on payment to workers for their shares, a scheme was proposed by the receiver to Robert Clark of Hill Samuel who sat on a committee representing creditors during the nationalisation process. The central premise being that the merchant banks would 'waive their deferred interest in exchange of an assignment of the worker's shares and the workers would be paid full face value, £1.89 million, out of the credit (£2.79 million) the waiver created in Rolls-Royce books'. The banks agreed and the workers were paid in full in September 1973. Ibid (*Nicholson E.R. (1993)* pg 49).

The receiver also stated that regarding the subcontractors that they devised schemes early on which provided cash advances in order to save their cash flow situations. When Rolls-Royce (1971) Ltd. insisted on renegotiating their contracts, many negotiated improvements in their terms 'which went some way to compensate them for the delay before their claims were paid in full'. Ibid *Nicholson (pg55).*

Rolls-Royce Crash: Key figures

Key figures: receiver Rupert Nicholson, company registrar Bill Bradley, joint liquidator Guy Parsons and Don Pepper, a board member at the time of the crash, pictured at a farewell staff dinner of Rolls-Royce Realisation Ltd in January 1983. Right, and on facing page, other faces involved in the crash, Bank of England governor Sir Leslie O'Brien, Prime Minister Edward Heath and Geoff Fawn, managing director of Rolls-Royce's aero engine division.

79. Farewell dinner; Rolls-Royce Realisation Ltd. in January 1983. Company registrar, B. Bradley; Rupert Nicholson and joint liquidator, Guy Parsons, Former Rolls-Royce board member D. Pepper.

COPYRIGHT ACKNOWLEDGEMENTS

CHAPTER 10

OUT OF THE ASHES – SIR KENNETH KEITH TAKES THE HELM

THE 1970's–1980's New foundations and consolidation

Following the shock of receivership in 1971, it was clear that it was going to take a long and sustained effort to turn Rolls-Royce's fortunes around and its financial recovery would require major changes to the company, its management structure and board membership. There was a need for new leadership and direction to take Rolls-Royce back to the forefront as a major aero engine manufacturer, capable of challenging and beating Pratt & Whitney and General Electric in the changing market for aircraft and engine technologies forecast for the 1970's and 1980's.

Leading From The Front – Sir Kenneth Keith

The man HMG chose to provide that leadership was Sir Kenneth Keith, who quickly attracted the nickname K^2. He was invited to take up the Chairmanship of Rolls-Royce (1971) Ltd. and did so in October 1972. This was a man with a larger-than-life personality who stood no nonsense or interference from government or from within the company. He possessed the strength of character and leadership qualities that were required to take the RB211 engine and Rolls-Royce back to the top. It was said he could be arrogant, especially when talking to ministers and ruthless when he needed to be to exert control in the company. He was also stated to be charming as well as having courage and had no qualms in networking at the highest levels when necessary, using all of his high-level contacts. He would ask various prime ministers of the day to speak to Presidents and prime ministers of other countries if Rolls-Royce needed to bring matters to their attention.

As a previous chairman of Hill Samuel, the banking group, Sir Kenneth was used to high finance, but had formed the view that £5 million was a lot of money. Sir Stanley Hooker recalled that, "We had added a '0' to his stature, because after a few weeks on the RB211 you soon came to understand that £50 million is peanuts". Ibid., (*Hooker S.* 1984).

Rolls-Royce for all intents and purposes continued to operate in its daily affairs as a publicly owned company whilst it was, of course, nationalised. This was based on a Memorandum of Understanding formulated in September 1971 with the Conservative government which set out the rights and obligations of each. The government was clear it did not want to be involved in the day-to-day management of Rolls-Royce's business but would expect to be kept informed of its forward planning and financial commitments. Rolls-Royce was to provide the Department of Trade & Industry (DTI) with full information about its operations.

Sir Kenneth wasted no time in reorganising the company, appointing Dennis Head as his new Managing Director and beginning to outline and set the company's targets and objectives for the immediate years ahead. Some of these were:

1. Maintaining its position as a competent designer and maker of aero engines.
2. Ensuring Rolls-Royce continued in the market for large civil engines and developing further derivatives of the RB211 engine family following on from the earlier RB211-22B and C engines, these being the RB211-524 & RB211-535.
3. Expanding and facilitating the company's small engine market portfolio and maximising new opportunities, particularly extending this to cover new industrial and marine applications.

Sir Kenneth had throughout his Chairmanship, an uneasy relationship with both Labour and Conservative governments having little time for ministers or civil servants. He often struggled with both the DTI and the National Enterprise Board (NEB).

By November 1972, Sir Kenneth stopped progress reports and planning documents being given to the DTI, which had been required as part of its monitoring function on behalf of HMG, because he himself doubted the accuracy of the information being given (something shared by the DTI who thought some of the forecasting data optimistic). The information was essential for the monitoring of Rolls-Royce's activities and the proper conduct of its business. Sir Kenneth wanted to effect internal reforms and improvements within the company and set about implementing these.

By October 1973, Sir Kenneth was able to provide the DTI with a general ten year and more detailed five-year plan. However internal systems were still prone to providing optimistic data especially with regard to future production costs.

October 1973 also saw war break out in the Middle East. The Arab/Israeli conflict resulted in major reductions in production of Middle Eastern oil supplies to the West and the world's industrialised countries. The effect of increased and heightened tensions caused the world's oil price to quadruple at one point. Additionally, the company was experiencing higher wage demands and encountered some industrial disruption to the business leading to increased operating costs.

The incoming 1974 Labour government placed Rolls-Royce in the hands of the NEB as its sole shareholder, the NEB having been introduced as an investment bank and major provider of money for development -taking major shareholdings in ailing and recovering companies and exercising a greater degree of control over them, with a view to hopefully, benefitting from their success in due course. This was much to the surprise of Sir Kenneth who could not see what role the NEB could play or in having a tripartite relationship with Rolls-Royce and the DTI. However, Sir Kenneth had been assured by the Prime Minister, Harold Wilson, that Rolls-Royce would remain directly under the control of the DTI.

A Memorandum of Agreement was negotiated between the Labour government, the DTI and the NEB (The DTI would be split into separate departments in March 1974, of Energy; Technology and Industry – but for this purpose, the relevant department was the Department of Industry (DoI)). The memorandum was broadly under the same terms as the previous memorandum published in September 1971, by the then Conservative government; Rolls Royce would have full autonomy and the NEB would act as intermediary providing money for development but also requiring approval of appointments to the Rolls-Royce Board as well as being informed of its forward planning, the company's performance and finances. Furthermore, Rolls-Royce were expected to inform and seek approval from the Board and the Secretary of State for Industry for projects exceeding £5 million (this sum was later increased). Sir Kenneth found the NEB role as intermediary in the company's decision making and monitoring particularly irksome in this tri-partite relationship and would later refer to the NEB as 'a bureaucratic contraceptive'. Cited in Ibid., Hayward, K. *(1983) (pg 206)*; *Pugh P. (2001) (pg 258)*.

Initially the Secretary of State for Industry was Tony Benn with whom Sir Kenneth did not get along and the feeling was mutual. Tony Benn had been elected to parliament representing Bristol and of course he had thousands of constituents who worked at the Rolls-Royce, Bristol site. At their initial meeting, on the 10th July 1974, to discuss the five year plan 1975-1979, it was made clear by Sir Keith, that due to a combination of pressures, such as :- inflation, oil prices, the economic crisis, including a three-day working week; the effect of government expenditure cuts and delay in support for the development of the RB211-524; reductions in military expenditure and the overarching delay in capitalising the new company, Rolls-Royce was forecast to face a negative cash flow of £100 million in 1975. The Government had not provided the assurances needed which would mean if not resolved, a repeat of the 1971 bankruptcy was, yet again, on the horizon.

It was clear that the company's financial position was still very precarious and that for the company to move forward with and develop further derivatives of the RB211, the RB211-524 &-535, it would require a substantial injection of new capital. This would become particularly important when in 1975 the world economy would enter into a major economic recession. Sir Keith wrote to the Governor of the Bank of England and the Prime Minister, Harold Wilson, to garner support. However, it was not until Tony Benn was replaced by Eric Varley in 1975, who had been Secretary of State for Energy in 1974-1975, and with whom Sir Kenneth seemed to have a more cordial relationship, that progress was made. He had previously written to him in October 1974, referring to the two five-year plans which he'd had prepared and submitted to HMG during his Chairmanship under the 1971 memorandum and complained about the lack of discussion in obtaining any commitment from government to investment in the company's future plans, which Sir Kenneth believed, was necessary for its survival as a commercial enterprise and technical force in the engine market. Soon after becoming Secretary of State for Industry, Eric Varley gave permission for Rolls-Royce to develop the RB211-524 engine, which would be certified at 50,000lb. thrust at the end of 1975. By extending further the engine thrust range from the original RB211-22 series to this capability, the RB211-524 engine would be an attractive proposition for later, long range versions of aircraft such as the TriStar L1011-500 and the Boeing 747.

RB211-524 Series Engines: 'A Bigger Beast' and the RB211-535 Series Engines: 'Smaller, Leaner, Cleaner'

The first variant of the Rolls-Royce RB211-524 engine was designated the RB211-524B, which entered service on the TriStar L1011-200 in May 1977 with the RB211-524B2 entering service in July 1977 on the Boeing B747-200. During the 1980's the engine went through several iterations, improving the thrust power which could be adapted for several different aircraft applications. The thrust moved from the original 50,000 lb. through a range of engine improvements to the RB211-524H variant operating at around 60,000 lb of thrust. This extended the aircraft's flying range along with the benefits of reduced fuel consumption. The RB211-524 engine's advantages and savings in improved fuel consumption, made it attractive to potential buyers at a time of worldwide high fuel costs and would eventually go on to win important orders for Rolls-Royce with Pan American World Airways (Pan Am) and Cathay Pacific, and eventually, after protracted and difficult negotiations between Rolls-Royce and British Airways, for its first Rolls-Royce powered, Boeing 747 aircraft, for which it was considered well matched (See Richard Turner's comment on this in Chapter 3).

The competition in the 1980's was fierce; the RB211-524 had proved itself, particularly in fuel advantage, but the worldwide recession in the early 1980's had the knock-on effect of reduced orders which meant the success of the RB211-524 could not be fully exploited. Ibid., *Pugh, P. (2001) (pg. 310).*

In addition, the decision to cease production of the TriStar by Lockheed left the -524 exposed to the competition with GE and P&W with their development of the CF6-50 and JT9D respectively, both in the process of being upgraded, challenging the -524's fuel advantage.

In the mid 1970's, Boeing had announced it was to build two new aircraft, the B757 & B767. Ernest Eltis and Geoffrey Wilde saw Sir Keith in London and persuaded him that the Company should develop the RB211-535 for the B757 aircraft. Sir Keith again wrote to Eric Varley outlining the importance of the project not only in terms of profit and export potential, but also for employment for the UK. He also pointed out the opportunity for Rolls-Royce to provide a launch engine on a Boeing aircraft for the first time in its history and believed the RB211-535 programme could produce £5 billion in total business. Sir Keith emphasised the survival of Rolls-Royce depended on securing the order with Boeing. Rolls-Royce was able to convince the NEB of its potential and in February 1977, Rolls-Royce was given the go ahead to build the RB211-535.

The RB211-535 engine was a scaled-down version of the earlier RB211-22B to power the new twin-engine Boeing 757 aircraft (a replacement for the ageing Boeing 727) and became a very successful seller for Rolls-Royce. With a smaller fan diameter, it was apparent that this twin-engine aircraft was designed to place a greater emphasis on engine reliability and lower maintenance costs with some basic redesigning of the engine architecture.

As the RB211-524 was considered a match for the Boeing 747, equally, the RB211-535 was considered a match for the Boeing 757 and its derivatives.

In December 1977, the financing of future development by means of launch aid, equity capital and loan finance was being discussed by the NEB and DoI. It became clear that financial commitments to develop both the RB211-524 & RB211-535 engines required just over £500 million (1978 prices) *(cited by Hayward K. (1983) pg.202). (The Comptroller and Auditor General's Report HC CAG 446(1979-80), para.15).* Launch aid would also be limited to the RB211 series, anything else would have to be financed by the company entirely and would cover a period 1978-1983. HMG would go on to provide a further

£180 million launch aid to Rolls-Royce in April 1980 and would waive its rights to the R&D levy on engines ordered in 1979.

In August 1978, British Airways had planned to announce orders for the Boeing 757 powered by the RB211-535C, which would enter service in January 1983 with Eastern and the improved RB211-535E4 in October 1984. However, as they had not obtained formal board approval, Boeing had to call an emergency Board meeting in September 1978, which then allowed Boeing to confirm at the Farnborough Air show, that its B757 aircraft would be powered by the RB211-535 engine. Rolls-Royce would go on to consolidate the RB211-535 as lead engine in the Boeing 757 aircraft.

As a launch engine in 1983, the RB211-535C variant operated around the 37,400 lb. thrust area and was later succeeded by the RB211-535E4 version at around a 40,100 lb. thrust level, both contained many of Rolls-Royce's most advanced technologies. These improvements included the world's first wide-cord fan blade, which is now a standard feature on all commercial turbofans, an integrated nozzle providing enhanced cruise efficiency and reduced noise characteristics. The introduction of a wide-cord fan blade (a hollow titanium blade made from two preformed pieces, bonded together around a honeycomb sandwich) was an engineering masterpiece. The manufacturing process was subsequently developed to produce this revolutionary fan blade by diffusion bonding, super plasticity forming the blade to its final shape. This required bonding three pieces of flat sheet titanium and then inflating the construction to expand the blade to its final shape in a single process. The improved RB211-535E4 won accolades from the world's airline industry for its reliability, maintainability and low operating costs. The RB211-535E4 would go on to become the engine of choice for over 80% of the Boeing 757 airline customers: Eastern Airlines eventually converted its whole fleet of aircraft to RB211-535E4 engines.

The following figures referenced in Ibid, *Pugh, P. (2001) (pg. 321)* of the actual spend in R&D, borne by Rolls-Royce, mostly taken up for the development of the RB211-535C, and the enhanced version, RB211-535E4, both developed for the 757 aircraft, were set out in the 'Long Form Report' drawn up by the Company auditors, Messrs. Coopers & Lybrand in detailing the Company's financial affairs ahead of the 1987 flotation. It was said this would increase from £62 million in 1981 for the RB211-535C to £131 million in 1982 and £131 million again in 1983. Any savings on R&D for the RB211-535E4, which commenced development in 1981 and would enter service in 1984, was not expected to reduce until 1984 and into 1985. This Report would go on to stress, yet again, the importance of the American market.

The Changing Of The Guard-Sir Francis McFadzean

The end of the 1970's decade saw a change of Chairmanship from Sir Kenneth, whose focus was more inclined towards sales, to Sir Francis McFadzean, later to become Lord McFadzean of Kelvinside in 1980. McFadzean was a friend of Sir Kenneth and in terms of his approach toward perceived interference from Government officials, held similar views. Sir Kenneth had announced his early retirement in November 1979, stepping down as Chairman in January 1980. In the same year the 1,000th RB211 engine had gone into production. However, between December 1982 and December 1984, only 126 new RB211 series engines would be delivered compared with the 350 envisaged in the previous year's plan.

Sir Francis McFadzean was to be well remembered for focusing on cutting costs, rather than marketing the engines and who it was said, only visited the Derby Headquarters on four occasions during his Chairmanship between 1980 and 1983. By the end of 1982 he oversaw a reduction in manpower; from 62,000 in 1980 (64,000 in 1971) to 46,000, and the streamlining of production operations by outsourcing production of small parts to external companies. This was set against the background of a worldwide recession in the early to mid-1980's, with inflation, fluctuating currency exchange rates, hugely increased oil prices, high interest rates and a consequent scarcity of airline orders for new aircraft. Against the backdrop of all of this and importantly with Lockheed terminating its L-1011 aircraft production by the end of 1982, Sir Francis McFadzean told the House of Commons, Trade and Industry committee that Lockheed's decision would mean a probable loss of turnover of approximately £250-£300 million in the late 1980's manifesting in a loss of profit between £50-£70 million. (Ibid. *Pugh, P.* (2001) pgs. 299-300).

It soon became apparent that the number one requirement for any engine manufacturer in this changed economic environment would be to concentrate on cost improvements, improving engine performance and reliability as well as extending component life on existing large turbofans, therefore resulting in better fuel economy and increased profits for airlines.

Sir Francis appeared before the House of Commons Public Accounts Committee on the 16th July 1980; Answering questions regarding the fixed contract rate with Boeing, inherited from the decisions taken by Sir Kenneth and his team, as not being hedged against a rise in sterling, (the primary exchange rate for Rolls-Royce engine contracts being the US Dollar), stated this was based in part on past experience.[16] He was talking about the original Lockheed

16 Sir Kenneth and his senior managers, having taken advice, based its contract with Boeing at $1.80 exchange rate crucially did not insure-hedge forward, against the impact of sterling strengthening against the dollar which began to rise in 1979. This was partly due to the UK benefitting from North Sea oil and also due to intervention

L1011 contract when the first 555 RB211-22 engines were expected to realise a loss of £65-£70 million but earned £54 million profit largely due to a currency swing in favour of a strengthening US Dollar against a slide in Sterling.[17] Neither the NEB nor the DoI were involved in discussions over the Boeing contract or the decision not to hedge against Sterling strengthening (although the NEB had endorsed the unanimous forecast from agencies that the trend of Sterling would be downwards) and and probably neither, he reported to the Committee, were they fully discussed by the full Rolls-Royce Board although indicating this was not unusual in large companies. *HC 779 Thirtieth Report, PAC (1979-80),Q (5179), 5656-62, (Appendix iv);* Cited in Ibid., *Hayward K. (1983), (pgs. 204-205).*

The Tay Engine

Following on from the challenges and turmoil of the 1970's, the first half of the 1980's remained financially difficult for Rolls-Royce, with a still detrimental exchange rate coupled with the fixed rate in the Boeing contract; however, things would improve in the second half of the decade. The 1980's saw Rolls-Royce developing a design derivative of the Spey 555, namely the Tay engine, part of Rolls-Royce's smaller engine portfolio, in order to match new aircraft opportunities appearing in this market sector. The new Tay turbofan engine, was wholly designed, developed and achieved engine certification at the East Kilbride site and this improvement on the Spey was later awarded the Queens award for technological achievement, a major success story for the employees at the Rolls-Royce, East Kilbride site, Scotland.

The Rolls-Royce Tay was successfully brought into service and found itself as the installed engine for the new Fokker 50 and Fokker 100 aircraft. It was also

---

by the Federal Reserve to 'cure' US inflation at the time by devaluing the dollar; the exchange rate of Sterling rising to $2.50 at one point, before it began to drift downwards in 1981 to $1.95 and returning to parity in 1985. The 'cure' hurt US exports amongst other things. Hedging against this would have fixed the risk to Rolls-Royce at $1.80, protecting its contracts against the rise in Sterling but not being able to take advantage should the exchange rate fall.

17 Regarding exchange rates adjustments and having experienced the devaluation of Sterling by Harold Wilson's Labour government in November 1967 and the various complex fiscal and monetary measures which accompanied it, Rolls-Royce was, "regarding provisions relating to adjustments to exchange rates, unwilling to agree to any mathematical formula for consequential price changes". It declared it would "adjust price quotations outstanding as well as prices payable for firm orders for supplies scheduled for delivery two years or more from the date of such further devaluation to take account of the devaluation, and all other circumstances and factors involved" Ibid *MacCrindle R.A., QC & Godfrey P.FCA(1973), (pg. 153).*

installed on many new variants of the Gulfstream corporate aircraft range, e.g., Gulfstream IV and V.

This again is an excellent example of Rolls-Royce benefitting from the engine family concept by way of a design derivative engine.

Collaborative Projects and changing Chairmen.

Rolls-Royce had a long history of forming alliances on aero engine projects and collaborative programmes both for military and civil engine applications. In the early 1960's this had included collaboration on the Tyne engine whereby Rolls-Royce allowed production of the Tyne under licence. The consortium included Hispano-Suiza- Spain (taken over by SNECMA–France in 1968), MAN-Germany and Fabrique Nationale-Belgium, known collectively as the "Tyne Consortium". Each company party to the licence built parts for itself and the partners: Rolls-Royce 20%, Hispano-Suiza; 44%, MAN; 28% and FN 8%. The final assembly was undertaken by both MAN and Hispano-Suiza. Extending into the 1970's the Adour engine with Turbomeca of France for the Anglo-French Jaguar aircraft. The late 1970's into the 1980's saw a military engine project where Rolls-Royce was involved with Germany and Italy, forming the collaborative company 'Turbo-Union' for the multi-role combat aircraft (MRCA), also known as the Tornado, powered by the Rolls-Royce RB199 engine.

'Necessity is the mother of strange bedfellows.'
Rolls-Royce and General Electric-Collaborators or Competitors?

Sir William Duncan on taking over the chairmanship of the Company in 1983, quickly became concerned about the company's ability to survive in continuing, difficult economic times. The decision by Lockheed to discontinue the L-1011 TriStar production programme in the early 1980's, due to falling sales, had led to its withdrawal from the civil aircraft market. It became more important that Rolls-Royce actively sought out and developed new income streams by way of participation in the new aircraft/engine collaboration projects now emerging. Against this background, discussions restarted as to a possible collaboration with Rolls-Royce's arch-rival and competitor General Electric. Intermittent discussions had taken place over the years covering a wide variety of projects, including the V-2500, that had at one point been considered.

Duncan commenced talks with Brian Rowe, President and CEO of General Electric, on a form of collaborative agreement in the big engine market. At the same time Rolls-Royce continued its development of the RB211–524 engine.

Understandably, GE viewed the Rolls-Royce engine as a serious threat and competitor to its own CF6–80 engine in the marketplace, for new aircraft orders.

Discussions continued on a risk and revenue sharing partnership (RRSP) between the companies with the aim of each company having an equal share participation in both the GE CF6-80 engine programme and the RB211-535 engine programme. Each would bring its own technical expertise and market opportunities to the table. It was clear from the outset of the agreement, which was signed on February 3rd, 1984, that this was not going to be an easy programme collaboration.

In 1985, the Chairmanship of Rolls-Royce had again changed. This time Sir Francis Tombs succeeded William Duncan who had died whilst in office on November 5th, 1984, taking over from the government's interim appointee, Sir Arnold Hall.

Rolls-Royce were advised by its American Lawyer, Phil Gilbert, that the agreement with GE couldn't continue in its current format and would need to be abandoned or revised. The basis of the advice was that Rolls-Royce could not be a collaborator and a competitor at the same time and expressed caution that the American Department of Justice may not accept that they could continue independently developing their own engines whilst collaborating on these same projects. In January 1986, Ralph Robins was suggesting an amendment to the Agreement which was concluded in the February. However, the successful bid by Rolls-Royce to supply British Airways with the RB211-524D4 engines for its new fleet of 747 aircraft, whilst collaborating on the GE CF6-80C2, finally resulted in the termination of the collaborative agreement with GE on November 19th, 1986.

> *"Throughout the period of collaboration Rolls-Royce had made a significant contribution to the GE CF6-80C2 programme assembling 12 engines and manufacturing 350 LP Turbine modules and HP compressor casings. On the other hand, GE made very little contribution to the RB211-535 programme assembling and testing 6 engines and subcontracting some of their share of the work back to Rolls-Royce." Ibid., Pugh P. (2001), (pg.319).*

Now, unencumbered by the restrictions of the collaboration agreement with GE, the development of the RB211-524D4 engine continued.

Pratt & Whitney, Man Turbo Union (MTU) and the Japanese Aero Engine Company (JAEC)

Another major, multinational collaboration programme began in the 1980's in which Rolls-Royce assumed a major partner leadership role for the V-2500 engine project, (which had previously been one of the potential collaborative projects considered with GE) for the Airbus A320 aircraft. In March 1983, Rolls-Royce entered into a partnership with Pratt & Whitney, MTU and JAEC which eventually became International Aero Engines. This collaborative joint venture set out to build an engine of around 25,000 lb. thrust for the then proposed 150-seat, single-aisle, Airbus A320. Later derivatives of this engine were also introduced on the McDonnell Douglas, MD-90 in 1995. Overall, in just over a decade, orders for and delivery of engines by Rolls-Royce totalled around 6,000, proving to be a very successful programme.

EDGING TOWARDS RE-PRIVATISATION

With an improving financial outlook from the mid 1980's, Rolls-Royce was steadily showing signs of financial stability following the restructuring, cost cutting, and manpower reductions of the previous five years. The worldwide recession of the early 1980's and the subsequent scarcity of significant airline engine orders was now starting to ease, the oil price had collapsed, there was a decline in inflation and better market prospects were envisaged up to the end of the decade.

Confidence had returned to the company and its board of directors, and a strong case was now being made for re-privatisation, to 'throw off the shackles' of government ownership so that Rolls-Royce could once again return to its pre-1971 status and position. Although high development costs had been incurred for the RB211-535C engine in these years, these were viewed as necessary to enable the company to meet the increasingly complex, technological advances in aero engine development.

Another positive factor was that Rolls-Royce had begun to reduce the technological risk and high cost of engine design and development, where practicable, by proving technologies in advance of the high expenditure phase of an engine project, as championed by Joe Clayton, Stewart Miller and Phil Ruffles, and by making technologies transferable between engines. By the mid 1980's, Rolls-Royce programme management had improved, to the extent that there were 'no delays in aircraft programmes since the Tri-Star programme in 1972' Ibid., *Ruffles P.(2014)*. Much had been learned from both Lockheed and Boeing during this time.

The company was now in a position where it could offer an engine within each principal sector of the small, intermediate and large aircraft markets, either alone, as in the case of the RB211 family of engines and the Tay, or in collaboration with other manufacturers, such as the International Aero Engines V2500 engine, which would enter service in 1989.

On April 28th, 1987, Rolls-Royce commenced the journey out of State ownership back into the private sector and was floated on the Stock Exchange. Known as 'impact day', on this date the price of ordinary shares that were to be offered to the public, and investors following 16 years of government ownership, was announced at an opening price of £1.70 per share, valuing the company at around £1.36 billion. In 1984 pre- tax profits were £26 million, and in 1985, £81 million. Sir Francis Tombs, company Chairman at the time, was able to say that the group had as a whole, earned a record pre-tax profit in the year to end 1986 of £120 million. There was an outstanding order book worth £3.1 billion, a broad product range with which to compete and improved financial controls and growth in productivity in recent years. There would be a further injection of cash from the government of £283 million.

> *"Since 1973, the company has operated on the basis of assurances that the Government would meet its debts in the extremely unlikely event of liquidation. As part of the arrangements for the sale, the company will receive an injection of new capital amounting to £283 million, a major proportion of which will be used to repay most of the company's borrowings. The prospectus notes that the board considers that, taking into account available facilities and the injection of new capital, the company will have sufficient working capital to meet present requirements. The Government assurances referred to will thus be withdrawn on completion of the offer for sale".*
>
> **The Secretary of State for Trade and Industry and President of the Board of Trade (Mr. Paul Channon), HC Deb Rolls-Royce [*Hansard*] Series 6 (28 April 1987) Vol 115 cc231-32**

Rolls-Royce -v- P&W

When American airlines placed an order with Boeing for 50 of their Boeing 757 aircraft in May 1988, an order worth $ 500 million with an option to purchase 50 more, potentially taking the order to a $1Billion, Rolls-Royce once again went into competition against an American Company, P&W, to supply Boeing with its RB211-535E4 engine against the P&W 2037 engine. An advantage of the RB211-535E4 was its improved performance, superior reliability, extended component life and also, importantly, lower noise; The engine satisfying Stage

3 of the FAA Noise Regulations, a bonus in enabling the engine to meet the noise constraint regulations imposed by night-time curfews on aircraft into some airports, such as Washington DC.

ENTER THE TRENT

The company had survived the severe recession in the worldwide aviation industry in the 1980's and now, as the new decade of the 1990's took hold, it saw itself poised for a period of further growth and the start of a new engine project. Under the influence and direction of Ralph Robins as Managing Director, the introduction of a new family of engines known as the Trent would see Rolls-Royce eventually establish itself as the number one manufacturer of large engines throughout the world and would become the mainstay of the company's income stream for many decades to follow.

The extract below, from Kevin Michael's book *AeroDynamic: Inside the High-Stakes Global Jetliner Ecosystem*, (2016) (Library of Flight). American Institute of Aeronautics and Astronautics, Inc. contains a quote by Sir Ralph Robins which may sound familiar, echoing Sir Denning Pearson's view many decades earlier on whether to proceed with the RB211 project.

> *By the late-1990s, Rolls-Royce was a transformed company. Under the leadership of Ralph Robins, former Rolls-Royce Chairman, it was a leaner and more global aero engine manufacturer boasting a broad range of turbofans, turboprops, and turboshafts. Against stern odds, Robins backed the Trent project against much larger rivals; he had played The Sporty Game, arguably betting the company on this vision, and was succeeding. Shortly after retirement, former Rolls-Royce marketing executive Robert Nuttall asked Robins about the difficulty of the Trent launch decision. His response was characteristically clear headed.*
>
> *'We had three choices. One, we don't do the Trent and the company fails; two, we try the Trent, it doesn't succeed, and the company fails; or three, we try the Trent, and we succeed. It was no decision at all–so here we are.'*

Without doubt, the decades of the 1970's/80's belonged to the RB211 engine and its derivatives and so accordingly, the decades of the 1990's and 2000's belonged to the Trent engine and its many developing variants which saw the entry into service of the Trent 500, 700, 800, 900 and Trent 1000 engines. Orders poured in for the Trent engine, times were good, order books were bulging, and the company's balance sheet looked very healthy–for now.

The Trent engine was to see Rolls-Royce eventually establish itself as the number one manufacturer of large engines throughout the world.

However, little would the company know then, that some of the very same Trent engines, so successful in their early days of operation, were to encounter major technical problems later in their life span that would cost Rolls-Royce many billions of pounds to modify and finally resolve.

Latest and largest – Trent XWB-97 (extra wide body)

"The Trent XWB is the world's most efficient large aero engine in service. Designed, engineered, and optimised for the A350 these perfect partners deliver 25% lower fuel burn and CO2 emissions. A clean start towards a more sustainable and profitable future". Boasting 97,000lb of thrust and 15% fuel improvement over the original Trent engine, the XWB-97 is set to save $6.4M per year, per aircraft on fuel alone. *(Rolls-Royce, News & Insights).*

In May 2022 Quantas placed orders for twelve Airbus A350-1000 aircraft powered by the Rolls-Royce Trent- XWB-97. "These aircraft will be capable of flying on the new, longest non-stop routes from London to Sydney and Melbourne. The engines are made at the Derby site which is set to build, service and maintain the engines. The machines can fly halfway around the world without needing to land, with the new London to Sydney route boasting a 20-hour flight time." *(Pegden T, Business Editor, Leicester Mercury, accessed via online edition of The Derby Telegraph, www.derbyshirelive.co.uk., Reach PLC (03/05/2022).*

In December 2023 Cathay Pacific placed an order for six A350F (freight) aircraft and also in the December, HMG announced, an historic deal between Turkish Airlines and Airbus to provide 220 aircraft, including 70 A350-1000 aircraft powered by the Trent-XWB-97 engine. Testament to the success of this latest derivative of the Trent engine.

ECHOES OF THE PAST AND BEWARE OF THE BLACK SWAN– COVID-19

History reminds us not to forget the cyclical nature of the aerospace industry- 'boom' or 'bust', let alone the financial recessions that take place also having an impact on its business. Rolls-Royce has surmounted many obstacles in over a century of operation. The 1971 bankruptcy of Rolls-Royce Limited and the 1986 decision by Rolls-Royce to go it alone and develop the Trent engine, could have seen the end of the company. Although technical problems were encountered the development of the Trent, like that of the RB211, led to the addition of several engine variants, all proving to be highly successful engines for many years.

However, prior to the unexpected shock of the COVID-19 pandemic and the almost catastrophic effects it has had on the aerospace industry, airlines and Rolls-Royce; the significant issues with the Trent 1000 engine, had to be resolved and were during Warren East's tenure as Chief Executive Officer, but at a huge cost. Several profit warnings over a 20-month period had to be given, the final blow being struck when the pandemic resulted, almost overnight, in the virtual grounding of all world airline travel. This had a devastating effect on the company's civil engine business, particularly as much of its mainstream income and profitability is linked to its 'TotalCare®' agreements, where its aftermarket, engine overhaul, spares and repairs, sales income, is generated. Simply put, reduced aircraft flying hours result in less demand for engine overhaul and spares.

Although Rolls-Royce remains tied to an airline industry which remains in trouble itself, still not having recovered to pre pandemic levels, its civil aerospace business still generates around 40 per cent of the group's underlying revenue.

Warren East had to take drastic action, resulting in an almost immediate manpower reduction of around 9,000 jobs worldwide which included the Derby site, and the sale of some of its secondary business assets, including ITP Aero in Spain in 2021. All of these actions emphasised the need to reduce expenditure and increase cash liquidity to strengthen the financial resilience of the company. In his 2020 address to shareholders, Warren East espoused the importance of treating the workforce with dignity and respect whilst making the tough decisions necessary.

> *"This is not a crisis of our making, but it is a crisis that we face, and we must deal with it. Our airline and airframe partners are having to adapt and so must we. Being told there is no longer a job for you is a terrible prospect and it is especially hard when all of us take so much pride in working for Rolls-Royce. But we must take difficult decisions to see our business through in these unprecedented times.*
>
> *Governments across the world are doing what they can to assist business in the short term, but we must respond to market conditions for the medium term until the world of aviation is flying again at scale. We have to do this right, which means we will work closely with our employees and trade union representatives as appropriate, look at any viable alternatives to mitigate the impact, consult with everyone and treat our employees with dignity and respect."*
>
> **Warren East: CEO(2015-2022) Rolls-Royce Holdings plc**
> **2020 Address to Rolls-Royce Shareholders (Extract)**

THE FUTURE-ULTRAFAN, ELECTRIC ENGINES AND SMALL MODULAR REACTORS

> *"What is happening is that the whole age of aeronautical development, culminating perhaps in the Dart engine, is coming to an end. We can no longer . . . just sit in our offices or go to the workshop and project our ideas in terms of more and more speed and more and more power. Modern technology in many areas other than aerospace now has to take a kind of U-turn. It has to meet human needs as distinct from mathematical and technological requirements.*
>
> *. . . In this engine we have an attempt, and I firmly believe, as there was no depression among the Rolls-Royce technicians and workers whom I recently met in the House, that is a successful attempt, to get the noise, the stink and the filth our of aero engines . . . I firmly believe that this will be a successful attempt. It it is successful, this will be not only a successful engine, but it will breed other engines, which is the characteristic of technological success in this respect. Personally, as I am much more of a political philosopher than a politician, I am rather more interested in the RB211 Mark IV and Mark V, the kind of engines which will be bred from this first experimental engine, than in the experimental engine itself."*
>
> **HC Deb, Mr. Raymond Fletcher MP Ilkeston. (26 February 1971). Rolls-Royce [Hansard]. (Vol 812, cc 1140-43)**

Whilst under the leadership of Warren East, Rolls-Royce announced some major progress on a new engine development that it claimed will redefine the world of jet engines. Future programmes and platforms see Rolls-Royce starting to build what is the world's largest aero engine, a demonstrator engine, UltraFan, to help redefine sustainable air travel for a new family of aero engines. Rolls-Royce sees a large part of its long-term future in the manufacture and success of the new UltraFan engine.

UltraFan involves journeying back into the past to the original RB211 design philosophy regarding the use of composite materials. These new engine composite parts include the manufacture of the engine's fan blades, using state of the art, fully automated construction methods at the new composite technology facility in Derby. These parts are made from carbon composite materials, stronger and lighter than conventional metals. There is no reason to suggest, therefore, that the return to using composite materials for fan blades will not be successful this time around given the advances in technology and are a far cry from the use of the ill-fated Hyfil fan blades of the early 1970's. It is claimed this new UltraFan engine will deliver significant weight, noise and fuel burn advantages to airlines.

A further major feature of this is the reduction of fuel consumption by a further 25% when compared with the Specific Fuel Consumption of the current Trent family of engines.

The continuance of this ethos of 'invent once, use many times' in the case of engine families, together with investment in developing innovative technologies will enable Rolls-Royce to maintain its leadership in aero engine design with the company's focus shifting on to environmental concerns.

To this end, having conducted and completed its first tests in early May 2023, using 100% Sustainable Aviation Fuel (SAF); on 28th November 2023, a test flight between London and New York successfully took place, using SAF to power the Trent 1000 engines on a Boeing 787, 'Dreamliner' aircraft. Rolls-Royce subsequently announced that it had completed compatibility testing, ensuring all of its civil aeroengines can operate wholly on SAF, guaranteeing net zero greenhouse gas emissions in support of both the company and HMG's policies on combating climate change.

Its pioneering work in partnership with other companies and with UK government backing in a project named ACCEL (Accelerating the Electrification of Flight), which aims to achieve decarbonisation of transportation systems and also move toward cleaner forms of flight using advanced battery and propulsion technology, will result in the company making a contribution towards tackling climate change and supporting Rolls-Royce's strategic ambitions in moving towards sustainable 'net zero' flights by 2050.

The ACCEL project's prototype battery-powered, single-seater plane, aptly named *Spirit of Innovation* has had a preliminary test flight of 15 minutes and was claimed to have broken the electric airspeed record by reaching 555.9 kilometres per hour over three kilometres. Its aim is to make zero emissions a reality and play an important part in the future of an Urban Air Mobility market going forward. The Company's new developments and proposals for electric engines are likely to constitute part of the company's engine portfolio in years to come.

It's a huge project to make air travel net zero and, as with all innovation, it is not without its risks, but Rolls-Royce is synonymous with outstanding engineering expertise, never standing still, always moving forward. It has all of its history to draw upon–its 'falls and rises'; its outstanding engineers, its collaborators, its managers and employees' contributions, past and present, all continuing to ensure the future of this outstanding company.

A modern-day collaboration in power generation -Small Modular Reactors SMR

As an example of a modern-day collaboration and innovation, the Rolls-Royce led SMR (Small modular reactors) consortium incorporates the best of nuclear engineering, construction, and infrastructure expertise. The collaboration project with Assystem, Atkins, BAM Nuttall, Jacobs, Laing O'Rourke, National Nuclear Laboratory, Nuclear Advanced Manufacturing Research Centre and TWI. In support of this initiative, Rolls-Royce announced in September 2022, the opening of its new Nuclear Skills Academy, supported by industry and education experts, employing 200 new apprentices "from all walks of life" to begin their training with the submarine arm of Rolls-Royce. Funding has been guaranteed to provide 200 apprentices each year with nuclear education across four different courses for at least the next 10 years. Apprenticeships are spread across all functions of the business, including engineering, manufacturing and business. They also provide fully funded further and higher education qualifications, while the apprentices receive hands-on practical work experience within the business. *Pegden T. et al, accessed via online edition of The Derby Telegraph, www.derbyshirelive.co.uk., (Reach PLC) (11/05/2022)*

The innovation from Rolls-Royce's engineers is without parallel and the exciting developing projects such as UltraFan and electric engines together with the announcements which came out of the COP26 UN Climate Change Conference held in Glasgow 2021, mean, as long as adequate government backing is received, development of small modular nuclear reactors (SMR) for power generation, could play an important role in the company through the coming decades. If sufficient Government funding comes through for the necessary R&D in the greener projects Rolls-Royce has invested in, it would assist in moving away from a portfolio dependent on the vagaries of oil. Investment again becoming relevant.

Made available from the Advance Nuclear Fund, HMG has so far committed up to £210 million with a further £280 million matched by private investors which has been used to build the business and to start the regulatory process to approve the reactor design and identify sites for the SMR plants.

As before however, there are other companies also interested in securing Government contracts in these greener projects, including one in which Bill Gates, founder of Microsoft and chair of TerraPower, GE-Hitachi (GEH)and Bechtel are involved. So once again Rolls-Royce finds itself competing with American companies.

BACK TO THE FUTURE – A NEW ERA DAWNS

Tufan Erginbilgic (CEO) took over from Warren East at the start of January 2023. drawn by the "strong brand, reputation, and its importance as a global company". Introducing himself at the 2022 full year results presentation at the stock exchange and online and acknowledging there were only "a handful of companies whose products have helped shape modern society, from the development of jet engine itself, to the wide range of engines and systems we make today"; He stated that Rolls-Royce has been underperforming for an extended period.

In a global address to staff, at the Derby HQ, in January 2023, he described the company as a "burning platform" and that "Every investment we make, we destroy value. . . financially underperforming every key competitor'. The company's performance was "unsustainable" and was at a level that could not continue. He believed that Covid had nothing to do with the current financial situation and Rolls-Royce had been underperforming for some time. "Given everything I know talking to investors, this is our last chance.".

Mr. Erginbilgic stated that Rolls-Royce's civil aerospace division, which builds and maintains large engines for widebody aircraft, was not generating cash or profit even before the pandemic. He warned that the company needed to change, and that no business generating low returns should be in its portfolio. Although stressing he was convinced that he could improve the company, he warned employees needed to urgently *"think differently, act differently, make a difference so this business corrects itself and we don't have much time"*.

Erginbilgic T.*; 'Winning Together', extract from First 'Town Hall' talk to Rolls-Royce Employees, January 2023.*

He later stated his motivation is "to get Rolls-Royce to a place where, in another 150 years, the company will still be doing great things for the world, as it has throughout its long history".

Tufan Erginbilgic profile page [Post]accessed 3rd February2023 (LinkedIn)Available from: https://www.linkedin.com/posts/tufanerginbilgic_last-week-i-talked-with-our-global-rolls-royce-activity-7026483565534044161-yY2-/?utm_source=share&utm_medium=member_desktop

The warning signs of further job cuts; between 2,000-2,500 roles potentially being removed globally, up to 6% of its 42,000 strong workforce.

"We are building a Rolls-Royce that is fit for the future. That means a more streamlined and efficient organisation that will deliver for our customers, partners and shareholders".

From the company's pioneering of workers shares in the 1920's, fast forward to August 2024 with Rolls-Royce announcing that it would, in September, gift 150 shares each valued at approximately £700 to its employees worldwide, which would be held for a minimum of three years. This follows the company posting strong first half year results of £1.1 billion almost double that of the same period in 2023.

Mr. Erginbilgic stated the company's transformation and progress had been possible 'thanks to the hard work and action of its people'.

L-1011
LOCKHEED

9 September, 1971

To: Peter Colston

From: A. M. Folden

Subject: Early Delivery of RB-211 Engines

I want to personally thank you for the excellent performance on the delivery of the RB-211 engines (6) to an ahead of schedule position.

I know you have worked hard and put in many hours trying to bring this program to fruition and to have it succeed must make you feel 9 feet tall.

Thanks, Peter, for a job well done!

Sincerely,

A. M. Folden

cc: G. Fawn

80. Lockheed letter of thanks to Peter Colston, Rolls-Royce, for engine delivery date improvements, September 9th, 1971.

81. Visit of Bob Fuhrman, Lockheed President, to Rolls-Royce, April 7th, 1972. Left to right: Eric Scarfe, Manufacturing Director, Bob Fuhrman, Peter Colston, RB211 Programme Director.

INFORMATION

Ref: D070422
April 7, 1972

LOCKHEED EXECUTIVE VISITS DERBY

Mr. R. A. (Bob) Fuhrman, President of the Lockheed California Company, builders of the L.1011 TriStar airliner visited the Derby Engine Division of Rolls-Royce (1971) Limited on Friday, April 7, for the latest information on progress with the RB211 engine which powers the aircraft.

Mr. Fuhrman toured the Production Test Facility, the Precision Casting Facility at Sinfin and the Main Works at Nightingale Road followed by discussions on the RB.211 programme status and product support status.

Pictured greeting Mr. Fuhrman (centre) are Mr. P.E. Scarfe, Divisional Director responsible for Manufacturing (left) and Mr. P. Colston, RB.211 Programme Manager.

END

Neg. No: W 87454

For further information :

Rolls-Royce (1971) Limited
P O Box 31, Derby
Telex: 37645

Enquiries to: J R Cownie

Phone - Office: Derby 42424
Ext 15

Home: Repton 2544

OR

Rolls-Royce (1971) Limited
14/15 Conduit Street,
London, W. 1
Telex: 263371

Enquiries to J Chappell or
M Jolley

Phone - Office 01 629 6201

Home: J Chappell - Northwood 21547
M Jolley - 01 722 5233

82. Rolls-Royce Information Bulletin following visit of Lockheed President, Bob Fuhrman, dated April 7th, 1972.

83. Factory tour–Visit to Rolls-Royce by families of Lockheed senior executives in 1972. Showing Denis Head (centre left), Eric Scarfe (centre right) and Peter Colston (rear centre).

84. Sir Kenneth Keith, Rolls-Royce Chairman (1972–1979).

85. Sir Kenneth Keith (centre).

13/7

ROLLS-ROYCE LIMITED
(In Voluntary Liquidation)

Regd. No. 87989 England
Regd. Office: Moor Lane, Derby.
Please reply to the Joint Liquidators at the address shown below.

Joint Liquidators: E. R. Nicholson, F.C.A.; Sir William Slimmings, C.B.E., C.A.; K. D. Wickenden, F.C.A.
Receiver: E. R. Nicholson, F.C.A. (Appointed 4th February, 1971)

YOUR REF: OUR REF: DH/LRD

D. Head, Esq.,
Managing Director,
Aero Engine Division,
Rolls-Royce (1971) Ltd.,
Derby.

11th July, 1973

Dear Dennis

Now that we have at long last resolved the most outstanding matter with Her Majesty's Government I think it is appropriate that at a more personal level I should like to say how much I enjoyed working with your colleagues and yourself, and my sincere thanks for the very valuable assistance you provided during those traumatic first few months of the Receivership and thereafter.

My own involvement with the Receivership, particularly as far as A.E.D., Derby is concerned, is now tapering off, but I must say that I found the past 2½ years has been a most challenging and rewarding assignment and made even more worthwhile in being privileged to work alongside competent executives.

I am certain that our projections for the successful future of RR(71) will be more than realised, even allowing for the fact that these projections formed part of our exercise in obtaining a good price for the business!

Yours sincerely,

D. Hilleard.
For and on behalf of
E.R. Nicholson
Receiver.

86. Receiver's letter of thanks to Denis Head, Rolls-Royce Managing Director, July 11th, 1973.

FIRST FLIGHT
BOEING 747
ROLLS ROYCE
SEPTEMBER 1976

This first-flight cover commemorates a renewed association among British Airways, Rolls-Royce, and The Boeing Company, highlighted by the first flight of a 747 powered by Rolls-Royce engines. That event took place today: the maiden flight of the first British Airways 747-200B. RB211-524 power plants, each rated at 50,000 pounds of thrust, powered the aircraft. Rolls-Royce provided Boeing with both the engines and the nacelles; British Aircraft Corporation, Weybridge, Surrey, built the engine struts.

Installation of the RB211 on the 747 will offer substantial improvement in the 747's payload range capability over that of the 735,000-pound gross weight 747-100 currently in service with British Airways. The RB211-powered 747-200B has a gross weight of 800,000 pounds, and with a full load of passengers and baggage, has a range of 5,200 nautical miles--450 more than the lighter weight model.

The aircraft now enters an intensive flight-testing program of about 160 flight hours. Certification by both the U.S. Federal Aviation Agency and the British Civil Aviation Agency is expected in April of 1977.

We are proud of our team's achieving this major milestone.

87. Commemorating the first flight of a British Airways RB211-powered Boeing 747 aircraft, September 1976.

88. Celebrating the Rolls-Royce and Lockheed L-1011-500 success. an advertisement placed in the Los Angeles Times on the occasion of the launch.

89. Rolls-Royce 1987, share flotation sale queues at the National Westminster Bank, Derby Marketplace.

90. Rolls-Royce 'Power and Prestige'.

91. Sir Ralph Robins, Rolls-Royce Chairman 1992–2003.

92. Sir Henry Royce statue outside Moor Lane, Derby:
Fly past by a Fokker 100 aircraft powered by the Tay engine.

93. Rolls-Royce Engine Fitters and Assembly Shop, Sinfin Site.

94. Rolls-Royce engines portfolio mid-1990s. Clockwise from bottom L to R; Tay; International Aero Engines, V2500; Trent 700; RB211-535E4; BMW/Rolls-Royce BR 710; Trent 800; Williams/Rolls FJ44; RB211-524G

COPYRIGHT ACKNOWLEDGEMENTS

Images 80–83 copyright Rolls-Royce Ltd. Supplied by Joanne Carlson from her father Peter Colston's papers.

Image 86 copyright receiver & joint liquidators with kind permission of the Rolls-Royce Heritage Trust

Image 87 copyright presumed Rolls-Royce Ltd. Supplied by Joanne Carlson from her father Peter Colston's papers.

Images 84-85, 88 & 90-94 all copyright Rolls-Royce Limited, with kind permission of Rolls-Royce Heritage Trust.

Image 89 copyright of *Derby Evening Telegraph* Newspaper Archives.

CHAPTER 11

FINDINGS AND REFLECTIONS ON THE ROLLS-ROYCE LIMITED RECEIVERSHIP PERIOD
by Former Senior Rolls-Royce Directors and Executives, Receiver and HMG Appointed Government Inspectors.

The official report of enquiry into the financial collapse of Rolls-Royce ordered by HMG gave a critical assessment of the high level of technical and commercial risk accepted by the board of directors in entering into the RB211 contract with Lockheed in March 1968. The history of the aerospace industry reminds us that there can be high risk and probability of failure associated with new aerospace projects, some of which have in the past been notorious for cost overruns and, in the worst-case scenario, project cancellation (as in the case of the TSR-2 fighter project in the 1960's).

In their investigation, the HMG inspectors had registered doubts as to whether the Rolls-Royce board had fully considered and evaluated all the risks and consequences of the RB211 project from the outset, including the possibility of failure. In fact, by sanctioning the continuing progression and commitment from the RB211-06 development programme in late 1968 to the RB211-22 programme early in 1970, the very action of the board increased the risk exposure of the company's existing net worth from 30% to 60%, based only on this single project. The acquisition of Bristol Siddeley Engines had an impact adding to Rolls-Royce's financial problems in subsequent years.

From contract commencement there was a constant drain on the company's resources resulting from the increasing financial demands of the RB211 project and along with it a growing cash liquidity problem. It was clear that accurate cost management of the RB211 contract and the control of escalating costs fast became unquantifiable, particularly as some major parts of the engine design were still incomplete at the time the contract was signed. With hindsight, it also seemed that any risk and consequence analysis of the contract terms appeared to have been overshadowed by the board's overarching desire to win and conclude an RB211 contract with a US aircraft manufacturer. At one point this resulted in the forecasted engine production price exceeding the contracted engine sale price, therefore putting the RB211 project immediately into a financial deficit position even before an engine had been delivered.

It was always going to be a gargantuan task to recover lost time in the programme delivery commitments to Lockheed alongside controlling the spiralling costs. However, in the end, the delay in delivery, remarkably, was only 4 months.

There's absolutely no doubt that following the bankruptcy, everyone involved with the RB211 programme worked incredibly hard to keep the project alive and performed exceptional efforts, which in some cases was to the detriment of their health, to ensure the future of Rolls-Royce.

'WHERE DID WE GO WRONG?' BY SIR DENNING PEARSON.

"1. *We were over-ambitious technically in what we could do in the timescale, Hyfil was one example and operating at higher temperatures was another.*
2. *We too easily agreed to the change to the method of mounting the engine when the programme was already underway.*
3. *We based our costs and timescale estimating methods for the RB211 engine on our previous experience with the civil Spey engine, which was not only completed on time, but also within the original development cost estimates.*
4. *We underestimated the indirect implications of the step up in the physical dimensions of the RB211 engine and the high initial rate of production.*
5. *Although we had escalation clauses in the production contract and allowances for escalation in the launching costs, the formula, in the case of the former, did not recover all the escalation although later we were able to improve the formula. In the case of the latter, the allowance was inadequate.*

Our financial problems stemmed from a culmination of the above difficulties on this contract. There are a number of other areas where we could and should have done better, as follows:

1. *Our accountancy systems did not readily provide cost information on a contract basis, particularly on cash flow. This information had traditionally been provided on a divisional basis. I was personally too tolerant of this for too long.*

2. *I failed to be bold enough in making organisational changes within the division early enough, as I was reluctant to disturb the group which was working so hard and enthusiastically on the RB211.*
3. *On a development project the spend is so determined by the demands of the project that there is only limited room for control of expenditure. We could have had better control with better data, although this, in my view, would only have had a second-order effect on actual expenditure.*
4. *It was, however, very difficult to predict with any accuracy the ultimate cost of the development and launching of the RB211 project at any point during the period when troubles were being encountered and corrected. The overlap of the development and production programmes on a new engine, which is normal in the aero engine business, is one of those difficulties as the excess production costs on the early production engines cannot be estimated without an engine specification which was changing as development proceeded. This was the problem IRC (Industrial Reorganisation Corporation) ran into, and which gave them the impression that we did not know what was going on. We should have been making forward financial forecasts on various assumptions about the outcome of various elements of the development, but this was not done.*

As Chief Executive I was concentrating all my efforts on supporting those engaged on the RB211 project as I judged at the time of the progress of this project was more important to the shareholders than anything else. I gave this my priority over my duties as Chairman of the Board."

Sir Denning Pearson (Former Rolls-Royce Limited Chairman)
from a paper dated November 23rd, 1971)

REFLECTIONS BY FORMER SENIOR ROLLS-ROYCE EXECUTIVES

SIR STANLEY HOOKER CBE, FRS, DPhil, BSc, FRAeS, MIMechE, FAAAS (Former Rolls-Royce Limited Technical Director, 1971)

Worth quoting again from Sir Stanley's book, Ibid *Not much of an Engineer* (1984)

"Between February and the autumn of 1971, the Derby works achieved a miracle in totally transforming the RB211 programme. The modifications needed extended to every part of the engine, but they were carried through with the speed and the spirit reminiscent of the Battle of Britain.

The costs incurred were controlled tightly by Ian Morrow, and rightly so because it was to be RB211 costs that had broken the original company. At the same time, I knew we were running a race, and it was frustrating perpetually to be told that a modification was not possible because the finance had not been cleared. My response was always the same, "Get on with it, and I will take personal responsibility for the costs". Such arguments took up a lot of my time and eventually I went to see Ian Morrow and said, "I am sure you are as keen as anyone to see the RB211 through to certification, but people are using your name to fight against every step we take. Please give me £1 million credit to use as and when I think fit". He readily agreed, and I never had to spend one penny of it, because the knowledge that it was there, enabled me to override any financial blockage, and ensure that no such blockages occurred."

JOE CLAYTON (FORMER ROLLS-ROYCE LIMITED CONTRACTS DIRECTOR)
EXTRACT FROM A PAPER 'CONTRACTS LESSONS' DATED 27th APRIL 1970

. . ."I conclude that in such a high-risk area as research and development it is quite unrealistic to believe that we can predict the future to the kind of accuracy necessary.

. . . fixed or maximum price type contracts are inappropriate to the high-risk R&D area.

. . . both the potential customer and the hopeful supplier cannot define what they want or can supply in the necessary depth or with sufficient foresight to allow a workable contract of sale to be drawn up at a sufficiently early stage.

. . . I do not think that we should delude ourselves that by clever contract drafting we can protect a situation that is not capable of protection.

The issues are wider than the contract processes, they are at the root of the need to manage the innovative process and create a profitable business.

. . . more emphasis should be placed on . . . independent development of critical components and on getting them quickly to test-the 'building brick' approach . . .

SIR ALEX SMITH (Former Head of Advanced Research at Rolls-Royce Limited, circa mid-1960s)

"My own assessment of the bankruptcy would have been very different, and I would have laid the blame for the collapse of Rolls-Royce firmly upon accountants, and upon the myopic accountancy procedures in the country. Year-to-year accountancy procedures are not relevant or appropriate for the design, development, manufacture and sale of aero engines. There, a cycle of some 20 years is needed, the cycle imposed by the requirements of engineering procedures, which are inexhaustible, rather than by the requirements of accountancy procedures which are made by accountants for the convenience of accountants. The design, development, manufacture and sale of aero engines is an ideal business in which Britain should be involved, for they have extremely high added value. The content of any engine is but a small amount of materials, and then an immense amount of brainpower, but the timescale for adding the brainpower must be measured in decades, not years."

Sir Alex Smith "*Lock up the swings on Sunday*" (1998) The Memoir Club; Bookcraft (Bath) Ltd.

THE RECEIVER, E. RUPERT NICHOLSON, summing up in his recollections Ibid. (1993) (pg. 55)

"I think the nation got value for money in Rolls-Royce (1971) Limited. Annual profit returns are misleading where research takes a decade and development costs billions. These finance matters need to be judged by decades and even then, remember that the real profit comes from the sales of spares and services."

HMG-APPOINTED GOVERNMENT INSPECTORS

EXTRACT FROM THE FINDINGS OF THE HMG APPOINTED INQUIRY INTO THE COLLAPSE OF ROLLS-ROYCE LIMITED

R. A. MacCRINDLE QC AND P. GODFREY, FCA, (PUBLISHED MAY 1973, *pg. 357*)

"At the end of the day the question remains, why did this great company end its days in receivership? Its main board included men who would've adorned any board room. The company was not lacking

in skilled engineers, it was not lacking in aero engine experience, it was not lacking in advice from outside advisors and consultants of the highest repute. It enjoyed the confidence of its bankers and of HMG. It was respected by its competitors; its products have an enviable reputation. Yet it crashed monumentally.

There can be no doubt that the effective cause of the crash was the drain upon the company's finances brought about by the RB211 project. To be sure, the company was not without other problems, for example, the merger with Bristol Siddeley some years earlier had increased its liquidity troubles, but in the event, this was not of relevant importance. The RB211 venture apart, none of the problems facing Rolls-Royce would have given rise, immediately or remotely to the burdens which submerged the company in February 1971. The size of the engineering task was seriously underestimated. From this basic fact flowed the appalling consequences recounted in the later enquiry and HMG report. Once the contract was signed, no system of financial control could have significantly influenced the course of events. We have been critical of the decision to conclude the contract. Yet although it failed to achieve its goals, in fairness it was also stated that in relation to the magnitude of the challenges involved, the performance of the company on the project thus undertaken was remarkable. The extent to which countless individuals contributed exceptional personal efforts to the task should not go unrecorded."

PASSING THE BATON

The position in which Rolls-Royce found itself in in the mid-1960's was one of stark choice, either to progress into the big engine market for its future, with all the inherent risks that came with that, or to refrain and in all probability face both a major contraction of the company's business operations or the possibility of a hostile takeover by one of its competitors, as highly unpalatable as it would have been then as it would be today.

Although some would say Rolls-Royce took a huge gamble in bidding and winning the RB211 contract (notwithstanding the severe consequences for several years that followed), it did so in the belief it would then establish a massive competitive advantage over its rivals in the growing large aero engine market at that time, which indeed it did by the end of the 1970's and into the next decade and beyond.

Rolls-Royce, as we know, survived the receivership. The RB211 technical problems were eventually resolved allowing it to go on to produce a very successful RB211 family of engines for different aircraft applications. The RB211 programme provided the company with invaluable programme knowledge and helped forge technological advantages over its rivals, particularly in the introduction of the wide chord fan blade for the RB211-535 engine. In addition, the platform created by the RB211 series of engines enabled the transfer of design and technological innovations to be incorporated into the development and production of the Trent series of engines which secured the company's business growth over more recent decades, albeit with further costs in rectifying problems over the past few years.

Any company that wishes to be at the forefront of technological change and development can't afford to stand still. With its focus on new product developments as well as engaging in environmental considerations, Rolls-Royce is looking to the future. However, with new projects R&D comes once again to the fore, requiring the full support and adequate government funding for the company to take full advantage, particularly with its development of SMR's where it faces stiff opposition not in the least from the US.

I return to the lead up to the Rolls-Royce receivership and what followed set out in this book, of the threads that run through it; the circular nature of history; the cyclical nature of the aviation industry; the growth of the company from local to global player, those leading it into the future and not least, the great skill, ingenuity and inventiveness of its workforce.

> *It is advisable that we should say – certainly my hon. Friends and many hon. Gentlemen opposite will share this view – that the Rolls-Royce company has now, as it had on 3rd February, many distinguished engineers and many fine projects and that the company has contributed, and will continue to contribute, to engineering standards of excellence in this country.*

Mr Phillip Whitehead MP Derby North.
HC Deb, Rolls-Royce [Hansard] (26 February 1971) (Vol 812 c1081)

> *"The 'Magic of a Name' is just as important as it ever was, but the 'Magic' will only be regained if the firm's future behaviour and engineering standing are restored and universally recognised by results".*

Air Commodore Francis Rodwell Banks (Rolls-Royce Adviser; Aerospace Engineer, Developer of the Internal Combustion Engine, Fuel Development Expert for the Rolls-Royce 'R' Engine (developed for the Schneider Trophy race) December 1971

"We need to earn the right to invest in our future so we can play a big role in the world, decarbonising the industries we operate in today. We are going to do this together, as one team, because I fundamentally believe history is written by the people who show up to shape things and to make a difference".

Erginbilgic T.; *'Winning Together', extract from First 'Town Hall' talk to Rolls-Royce Employees, January 2023.*

https://www.derbytelegraph.co.uk/news/derby-news/new-rolls-royce-boss-tufan-8081414

Tufan Erginbilgic profile page [Post]accessed 3rd February 2023 (LinkedIn) Available from:

https://www.linkedin.com/posts/tufanerginbilgic_last-week-i-talked-with-our-global-rolls-royce-activity-7026483565534044161-yY2-/?utm_source=share&utm_medium=member_desktop

SOURCES OF RESEARCH AND BIBLIOGRAPHY

During the preparation and writing of this book, the author has researched, consulted or quoted from the sources listed below in relation to the receivership and any other relevant material.

REPORTS AND PAPERS

Department of Trade and Industry (1972). Rolls-Royce Ltd. and the RB211 aero engine (Cmnd 4860). HMSO.

MacCrindle, R. A. and Godfrey, P. (1973). Rolls-Royce Limited; Investigation under Section 165(a)(i) of the Companies Act 1948. HMSO.

Plowden, Lord (1964-65). The Report of the Committee of Inquiry into the Aircraft Industry (Cmnd 2853). HMSO.

PUBLICATIONS

Gunsten, Bill (1971). The phoenix of Derby. *New Scientist*, 9 December.

Owen, Kenneth (1974). RB211 growing pains of a special engine. *Design Magazine* (306), 72–75

The Economist (1971) – Issues 13.02.1971; 13 .03.1971; 24.04.1971; 01.05.1971; 12.06.1971; 22.01.1972; 31.03.1973.

ROLLS-ROYCE HERITAGE TRUST ARCHIVE MATERIAL

Pearson, James Denning, Sir (1964). The Development and Organisation of Rolls-Royce Limited. Paper written & presented to the London School of Economics, November.

Tombs, Francis, Sir (1991). A History of Rolls-Royce. GKN Lecture

Miscellaneous documents

BOOKS, BROCHURES, NOTES, EXTRACTS AND JOURNALS

Cownie, J.R. (1990). Rolls-Royce's Big Fan. *Civil Aviation Review, London* Aircraft Illustrated, Ian Allan Publishing.

Donne, Michael (1981). *Rolls-Royce, Leader of the Skies.* Frederick Muller.

Gray, Roberts (1971). *Rolls on the Rocks: The History of Rolls-Royce.* First published by the Compton Press and Panther Books Ltd (simultaneous edition)

Hayward, K. (1983) *Government and British civil aerospace*: A case study in post-war technology policy. Manchester University Press.

Hooker, Stanley, Sir (1984). *Not Much of an Engineer: An Autobiography.* Airlife Publishing.

Ingells, Douglas J. (1973). *L-1011 TriStar and the Lockheed Story.* Aero Publishers.

Michaels, Kevin (2016). *AeroDynamic: Inside the High-Stakes Global Jetliner Ecosystem* (Library of Flight). American Institute of Aeronautics and Astronautics, Inc.

Nicholson, Rupert (1993). *Rolls-Royce Limited: Recollections of the Receiver and Manager.* KPMG Peat Marwick.

Porter, Andrew (2013). *Transatlantic Betrayal: The RB211 and the Demise of Rolls-Royce Ltd.* Amberley.

Pugh, Peter (2001). *The Magic of a Name: the Rolls-Royce Story*, Part 2, Icon Books.

Reed, Arthur (1973). *Britain's Aircraft Industry: What went Right? What went Wrong?* Dent.

Rolls-Royce Heritage Trust. Exhibition Guide Notes.

Rolls-Royce plc (1986). *The Jet Engine.* Rolls-Royce plc.

Rolls-Royce (1968). Supplier Conference brochure.

Ruffles, Philip C. (2014). *The History of the Rolls-Royce RB211 Turbofan Engine* (Historical Series No. 47). The Rolls-Royce Heritage Trust.

Smith, Alex, Sir (1998). *Lock up the swings on Sundays.* Memoir Club.

Goddard, Ken (2000). *The Rolls-Royce Tay Engine and the BAC One-Eleven* (Historical Series No. 30). The Rolls-Royce Heritage Trust.

VIDEO/AUDIO

Lockheed Aircraft Corporation video '*Lockheed L-1011 TriStar – Too Advanced for us.* Available on You Tube. https://youtu.be/dAjezUfdBXE?feature=shared

Norman Innes, BBC Radio 4 "The Crash of Rolls-Royce", broadcast on 23.01.1972 (BBC Radio Derby's sound archive).

PERSONAL INTERVIEWS

Please refer to Acknowledgements.

Guy Parsons (PMM/KPMG). My Life's Story and Memoirs (unpublished).

The late Peter Pugh

The late Mike Kerr

AEROSPACE AVIATION MAGAZINES

Flight International

Aviation Week

NEWSPAPERS – NATIONAL

The Financial Times

The Daily Telegraph

The Sunday Times

The Observer

The Independent

NEWSPAPERS – Local

Derby Evening Telegraph, plus special *Derby Evening Telegraph* editions and supplements on Rolls-Royce Ltd

Leicester Mercury

EX-EMPLOYEES' ASSOCIATIONS

Lockheed Martin Star Dusters (Lockheed retired and former employees' association)

Rolls-Royce Retired Employees Association

PROFESSIONAL SOCIETIES

Royal Aeronautical Society

PERSONAL CONTACTS

Former Rolls-Royce work colleagues

ACADEMIC PAPERS

Lazonick, W. and Principe, Andrea (2004). Sustained Innovation: Career Engineers, Stock Markets and the Theory of the Innovative Enterprise. *SPRU Working Paper Series 121*, Science Policy Research Unit, University of Sussex Business School.

Norwood B. 'The Kennedy Round: A try at a linear trade negotiation', *The Journal of Law and Economics*, Vol.12, No.2, Oct.1969, pages 297-319. (University of Chicago Press Journals).

LIBRARIES AND ARCHIVES

Rolls-Royce Heritage Trust Archives

Derbyshire Records Office

Derby Local Studies Library

University of Derby Library

The National Aerospace Hall of Fame Archive (USA)

Hansard

House of Commons Parliamentary Archive

House of Commons Parliamentary Papers Online © 2007 ProQuest Information and Learning Company

Dictionary of Irish Biography; (https://www.dib.ie/index.php/biography/huddie-sir-david-patrick-a4128)

(https://creativecommons.org/licenses/by/4.0/legalcode).

Imperial College Archive

Institute of Mechanical Engineers Archive

HERITAGE SOCIETIES

Rolls-Royce Heritage Trust

Sir Henry Royce Memorial Foundation

British Airways Heritage Collection

ROLLS-ROYCE LIMITED

Various documents and extracts from Rolls-Royce in-house magazines and newspapers (*Rolls-Royce News*, *Reporter*, *World*) and Rolls-Royce technical and sales publications and press releases

SOCIAL MEDIA NETWORK GROUPS

LinkedIn. The Rolls-Royce Professional Network

Various Facebook interest groups: including Derby of Yesteryear, Lockheed L-1011 TriStar

MEDIA AND COMMUNICATIONS

Radio Derby news archives

Derby Evening Telegraph photography archives and the *Derbyshire Business Live* website; derbyshirelive.co.uk

COMPANIES CONTACTED

Rolls-Royce Limited

Lockheed Martin Corporation

KPMG

General Electric Company

ACKNOWLEDGEMENTS

This book project has taken me on an amazing journey over the past few years. It is only fitting therefore that I recognise and acknowledge all the people that helped me to make this, my first book, possible, no matter how large or small their contributions have been.

Firstly, I would like to extend my special thanks and appreciation to Rolls-Royce Limited and the Rolls-Royce Heritage Trust for kindly granting me where pertinent, the copyright permissions to utilise selected photographic images, newsletters and technical publications, documents, and quotation extracts which I have incorporated into the text.

My special thanks are also extended to the following people: Neil Chattle former chief executive and Peter Collins, former director and secretary of the Rolls-Royce Heritage Trust and Mike Stewart, Archivist of the Trust who all provided me with support and assistance in response to my many questions and allowed me research access to the Trust's Archives.

Phil Ruffles and John Coplin for their expertise and answering, with patience, my questions clarifying technical information .

Former Lockheed Aircraft Corporation employees, particularly Robert Paul and acknowledging the late Mike Kerr.

It would also only be right to record here the help and assistance given to me by Mrs. Brenda Wright, Rupert Nicholson's daughter, and also the late Guy Parsons, a close senior associate of Rupert Nicholson's 1971 receivership team and joint liquidator.

My gratitude also to the following for their exceptional assistance in lending and giving permission to use many personal photographs, documents, articles and contributions and their help in locating some of the key people for this project. They are listed in no particular order.

Bob Huddie

Philip Parsons

Philip Hall, The Sir Henry Royce Memorial Foundation

John Coplin

Richard Turner

The late Dr Ian Macpherson

The late Tony Lintott

Kevin Miller

The late Bill Grattidge

Photography advice, images and photographs sourced.

Rod Jewell (Rod Jewell Collection)

Geoffrey Thomas

Derby Local Studies Archive Photographs (formerly *Derby Evening Telegraph* photograph archives)

Simon Pank

BBC Radio Derby

Clare Cully

Andy Twigge

Formerly of BBC East Midlands Today

Mike O'Sullivan

The *Derby Evening Telegraph*

Jane Goddard

Derby Local Studies Library

Mark Young

Linda Webster

Cartoon Provision

Graham Newton

Lockheed Martin Star Dusters (ex-employees' association)

John Plowman

James Fitzpatrick

Sherm Mullin

John Shelton

Book Foreword

Phil Ruffles

I would also like to extend my thanks and appreciation to all contributors and the many people who gave me helpful assistance and encouragement in researching and writing this book, including:

The Rt. Hon. The Lord Tebbit of Chingford CH,PC, Sue Allen, Nick Tarr, David Whetton, John Bull, Ian Anderson, Dick West, David Marshall, Lynn McMurray, Brian Wood, Angie Paige, Gurmit Singh, Maureen Underwood, Graham Newton, Bill Meehan, Roger Taylor, Paul Conaghan, Alan Buxton, Roger Hollick, Dave Parfitt, Bev Ridgeway, John Plowman, James Fitzgerald, John Beddoes, Lynn McDonald, David Daykin, Giles Cattermole, Charles and Margaret Cuddington, Roger Hill, Barry Sutcliffe, Sheila Furness, Len Matthews, Roger Gregory, Barry New, Beth Nielsen, Simon Wilson, David Morris, Alan Spray, Kevin Miller, David Reade, Ian Paige, Tony Holme, Peter Craig, Jim Clinton, Dave Rosser, Eric and Gwyn Lancashire, Alastair Hunt, Mick Walker, Graham Malley, Steve Shepherd, Mick Lomax, Dave Tomlinson, Joanne Carlson and Susan Watts, George Manzor, Ian Tucker, Manuel Duarte, John Haynes, Malcolm Cresswell.

Apologies to anyone I may have inadvertently omitted.

Finally, last but not least, a heartfelt thank you to my wife, Barbara for her support, critical eye and invaluable manuscript assistance along the way.